ADVANCED NANOMEDICINE AND NANOBIOTECHNOLOGY

ADVANCED NANOMEDICINE AND NANOBIOTECHNOLOGY

Viroj Wiwanitkit

Nova Science Publishers, Inc.
New York

LIBRARY OF CONGRESS CATALOGING-IN-PUBLICATION DATA

Viroj Wiwanitkit.
Advanced nanomedicine and nanobiotechnology / Viroj Wiwanitkit.
p. ; cm.
Includes bibliographical references.
ISBN 978-1-60456-435-8 (hardcover)
1. Nanomedicine. 2. Nanotechnology. 3. Biotechnology. I. Title.
[DNLM: 1. Nanomedicine. 2. Nanostructures--therapeutic use. 3. Nanotechnology. QT 36.5 V819a 2008]
R857.N34V57 2008
610.28'4--dc22

2008005517

Published by Nova Science Publishers, Inc. ✣ New York

CONTENTS

Chapter 1

BASIC PRINCIPLES OF NANOSCIENCE

INTRODUCTION TO NANOSCIENCE [1 – 7]

Nanotechnology is an applied technology relating to the management, construction, and synthesis of devices at the atomic or molecular level that possess the sizes between 1 and 100 nanometers. The nanodevice has its special function in physical, chemical, and biological aspects which makes it useful. Nanotechnology is the applied usage of knowledge on nanoscience [8 - 10]. Nanoscience is a new science focusing on the natural aspects of nanolevel conditions. Nanoscience covers three main scientific areas: physical, chemical, and biological sciences.

The main question for the beginner in nanoscience is what the nanolevel is. “Nano” is a scientific term describing the 10^{-9} level. The nanolevel is smaller or fewer compared with the well-know microlevel (micro = 10^{-6})

Nanoscience covers areas of dimension. Length, as a one-dimensional aspect, is well-described in nanoscience. Measurement in nanometer (nm) is generally used in nanoscience. In addition to length, area as a two-dimensional aspect and volume as a three-dimensional aspect are also focuses in nanoscience. Measurement of small volume as nanolitres (nL) is also generally used in biological and medical science.

Table 1. Levels in science

Level	description
Kilo	10^{3}
Deca	10
Deci	10^{-1}
Milli	10^{-3}
Micro	10^{-6}
Nano	10^{-9}

Table 2. Examples of comparison of objects in different lengths

Size in length (nm)	Examples
10^8	Small ball
10^6	Insect
10^4	Cancer cell
10^3	Bacteria
10^2	Virus
10	Antibody
1	Glucose
10^{-1}	Hydro

History of Nanotechnology and Nanoscience [11 - 13]

It can be said that nanotechnology and nanoscience are not actually new things in the scientific world. In the past 50 years, a blooming of immunology led to many new challenges. Continuous developments led us to the years of molecular biology and genomics, which is a phase of microscience and microtechnology. However, the completion of the genome project led us to the post-genomic era. New science emerged to deal with bio-chemo infirmities [14 - 20]. In addition, a new development known as nanoscience and nanobiotechnology was also launched.

At present, nanoscience and nanotechnology focus on objects ranging in size from 1 nm to 100 nm. At these nanolevels, objects have new, different functions from their general functions at the gross level. The advent of nanoscience and nanotechnology helps us control the accuracy and precision of new nanoobject usage.

Of interest, nanoobjects are not new things. Human beings have used nanoobjects for a long time. In the West, the Lycurgus cup, a magic golden cup which could automatically change its color, made in A.D. 300 [21 - 23], is the best example. The previous way which was used to grind Roman diatreta glasses was later elucidated with the aid of reproductions of both the network diatreta from Cologne-Braunsfield and Niederemmel and the Lycurgus beaker [23]. Josef said that first it was shown how to copy the grinding characteristic features which were closely connected with stone cutting but without relation to the work of Roman glass blowers, glass cutters, or engravers [23]. Josef also mentioned that the comparison of the network diatreta which was made based on mathematical and technical principles, with the Lycurgus cup presented a completely different handling of the grinding wheel [23].

Table 3. New functions or properties of objects at the nanolevel

Properties	Description
1. quantum effect	change in color, electrical property, magnetic property
2. surface effect	increased reaction due to increased surface area
3. interface effect	change due to scaling laws

Hornyak et al also reported that transmission color at specific wavelength of red-orange with maximum lambda in the range 512 to 519 nm was necessary to obtain the scarlet-red exhibited by the famous Lycurgus cup of 4th century Rome [24]. In the East, gold and silver leaflets were also produced and used in ancient Chinese, Indian and Siam dynasties. This refers to the fact that we had known and used nanoobjects for a long time, but we have just recently focused our interest on these nano-level objects.

PRINCIPLE OF SIZE DIFFERENCE

Perception of size is a basic perception for a human being. "Size" is one of the knowledge targets for children in kindergarten. The relationship between long and short, big and small, and heavy and light are basic fundamental skills that humans have to develop. This can be a very basic principle for studying nanoscience. It can be used to explain a complex natural phenomenon. Examples of the usage of the principle of size difference to explain natural phenomena are hereby given to help readers better realize the basic principles of size difference.

A. An Appraisal of Blood-Borne Viral Germs in Breastmilk and the Feasibility of Passing through the Enterocytes of an Infant

Introduction

The benefits of breast milk have been well delineated, and breastfeeding should be encouraged whenever probable, however, the potential transmission of blood-borne germs warrant the recommendation that women in indusexperimentized countries who are infected with some of these viruses not breast feed their infants [25]. Many maternal diseases caused by bacteria, virus, fungi, and parasites might sometimes be transmitted via human milk [26]. Previous reports have pointed out that mothers infected with HIV and T-lymphotropic human viruses (type I) should also not breastfeed [26]. With many other infectious diseases, a careful approach should be taken, but in general breastfeeding is maintained [26].

In countries with a high prevalence of blood-borne germ infection, transmission to infants commonly occurs from an infected mother around the time of delivery; another mode of transmission is from child to child. However, concerns still remain on the issue of transmission via breastfeeding. Here, the author analyzed the feasibility of passing through the enterocytes of some blood-borne viral germs (hepatitis C virus, hepatitis D virus, hepatitis SEN virus, hepatitis TT virus, Cytomegalovirus and Epstein-Barr virus) in breast milk based on considerations at the nanostructure level.

Materials and Methods

Pathophysiologically, the two main reasons for the transmission of any germs to a neonate via breastfeeding are a) the germ is secreted out from the mother via breast milk and b) the germ can pass through the alimentary tract mucosa of the neonate entering into their blood streams. An important indicator for the possibility in tranferring the germ is the relative size of the germ compared to the pore (at the maternal mammary gland and at the neonate alimentary tract mucosal epithelial junction). Here, the author reviewed the previous information on these topics. Further comparison to the pore size is performed. Here, the author selected some important blood-borne germs for further study. The selected germs included hepatitis C virus, hepatitis D virus, hepatitis SEN virus, hepatitis TT virus, Cytomegalovirus (CMV), and Epstein-Barr virus (EBV).

Results

According to the review, the maternal mamary gland pore size is about 200 – 400 nm [27] and the neonatal intestinal epithelium junction pore size is about 0.4-nm in diameter [28]. The reported sizes of each virus [29 – 34], with comparison to the pore sizes, are presented in Table 4.

Discussion

International agencies recommend exclusive breast-feeding for 4-6 months followed by continued partial breast-feeding into the second year of life in order to bring good infant and child health and decrease the damage caused by the malnutrition-infection cycle [34]. Breast-feeding is believed to benefit infants because breast milk consist of the ideal mixture of nutrients for infants, because it consist of factors which help in the development of the infant's gut and immune system which prevent germ invasion, and because exclusive breast-feeding prevents the intake of germs in food or water [34]. However, breast milk itself may consist of bacteria or viruses and this problem has come to the forefront with the human immunodeficiency virus (HIV) epidemic, since it is clear that breast-feeding is an important mode of mother-to-child transmission of HIV. The present article reports the examination of the feasibility of passing of some important blood-borne viral germs in breast milk through the enterocytes.

The hypothesis is based in part on the notion that hepatitis C virus, hepatitis D virus, and EBV are in human milk [35 – 38]. Yet, for other viruses, there is little published evidence that those living viruses are in human milk.

Table 4. Reported viral sizes and comparison to the pore sizes.

Virus	Size (nm)	Comparison to the pore sizes	
		Mammary gland pore	Gut epithelium junction pore
hepatitis C virus [29]	60	Smaller	Larger
hepatitis D virus [30]	32	Smaller	Larger
hepatitis SEN virus [31] hepatitis TT	26	Smaller	Larger
virus [32]	30	Smaller	Larger
Cytomegalovirus [33]	150	Smaller	Larger
Epstein-Barr virus [34]	130	Smaller	Larger

Although it is not proven, the possibility of all studied viruses to be transmitted via breastmilk can be expected. However, according to this study, none of the studied viruses seem unable to directly enter into the pore of the neonate enterocyte. This is warranted by the fact that although virus is often demonstrated in breast milk, its transmission is extremely rare [38].

In addition, there has been no report of transmissions of the studied viruses via breastfeeding. Based on the particle size as a single factor, the author hereby proposes that transmission on the studied viruses by breastfeeding is difficult. It can also be concluded that that the "pore size" in the maternal mammary gland and in neonatal intestinal epithelium influence infectivity. The impairment of the anatomical structures in the preterm might be another risk factor for obtainting an infection via breastfeeding. However, the probable infection due to other processes, such as contact with maternal body fluid consist of viruses during daily contact, might be a more important route of transmission. Nevertheless, there are other factors that should be considered in the feasibility of transmission of the studied viruses via breastfeeding.

First it should be noted that passing via pore is only the probable mechanism of viral transmission. Of interest, there is no previous report that the size of the virus or the diameters of the pore of epithelial cell junctures are completely responsible for these viral infections. In addition, sometimes the viral receptor is another considerable factor. There is evidence that many viruses gained entrance to mucosal cells by specific receptors, such as CD155 for polioviruses [39]. Some different size viruses, for example adenoviruses are large and polioviruses are small, yet both infect the intestinal tract. Also, variola major virus is exceptionally large but readily infects the upper respiratory tract [40]. Indeed, hepatitis C virus entries into cells appear to occur via CD81 and/or the class B scavenger receptor [41], hepatitis B and hepatitis D appear to have cellular receptors [42], EBV entries into cells appears to occur via membrane fusion [43], and CMV viral envelope proteins appear to interact with cellular receptors to get cell entry [44]. Second, even though cytomegalovirus [45 – 48] and hepatitis TT virus [49] are in human milk, they do not infect the recipient infant. The converse is the case. Those live viruses in human milk infect the nursing infant, but the infant does not develop the disease because of the infections [45 – 49]. In fact there is little evidence that hepatitis TT virus is a germ. Finally, defense mechanism is another considerable factor. There are a host of defense mechanisms that determine whether a given virus infects a mucosal site. These include mechanical barriers such as mucins, specific secretory antibodies, antiviral lipids, interferons, complement components [50], macrophages, and cytotoxic T cells. Even if the size of the virus and the size of the pores of epithelial junctures play a role in infectivity, they would most likely be over-shadowed by these proven defense mechanisms. However, in infectious cases, use of breast milk should be completely controlled, with strict attention to infection control issues in obtaining, storing, and processing the milk [38]. Physicians should be aware of the risks of transmission of viral germs with fresh maternal breastmilk [38].

B. Importance of the Size of Immune Complexes in Glomerulonephritides

Introduction

The passive trapping of preformed immune complexes is responsible for some forms of glomerulonephritis that are related with mesangial or subendothelial deposits [51]. The biochemical characteristics of circulating antigens play important roles in determining the biologic activity of immune complexes in these cases [51]. The immune complex plays an important role in the germesis of many glomerulonephritis diseases especially for lupus nephritis, acute post streptococcal glomerulonephritis (APGN) and IgA nephropathy.

The activation of the complement system is believed to be an important factor leading to the destruction of glomerulus [52]. However, the actual germesis of the glomerulonephritis has never been clarified. The structures of some immune complexes have been documented in the recent year. However, the knowledge of the effect of the size of immune complex on the glomerulus is not well documented. Here, the author performs an analysis on the size of the immune complex in important immune glomerulonephritis diseases to appraise the effect of the size of the immune complex on the glomerulus.

Materials and Methods

The database RCSB PDB Protein Databank [53] and PubMed (www.pubmed.com) was used for data mining of the molecular structure of immune complexes in lupus nephritis, APGN and IgA nephropathy. The searched results were further processed to be in the format of three - dimension (3D) molecular structure by Swiss-Pdb Viewer (GlaxoSmithKline Rand D and the Swiss Institute of Bioinformatics). The geometry or shape parameters of the complexes were then determined.

Results

The output data for immune complexes in lupus nephritis (1 YTY), APGN (1KTK) and IgA (10W0) nephropathy are derived. Using Swiss-Pdb Viewer, the shape-parameters of the three complexes are detected and presented in Table 5.

Discussion

Immune complex generation is a basic finding in many glomerulonephritis diseases. This is well-known as the hypersensitivity type 3. The entrapment of the immune complex in the glomerulus is believed to be important in germesis of glomerulonephritis. Immune glomerulonephritis initially develops as a result of an inappropriate immune response, either inadequately directed to self-antigens, or the inability to effectively eliminate a foreign antigen [54]. This abnormal immune response leads to the formation of circulating immune complexes, auto-antibodies or self-reactive effector T cells that target glomerular antigens, or to the inappropriate regulation of IgA producing lymphocytes (Berger disease) [54]. In most of the cases, the initial immune injury induces an inflammatory response [54].

Table 5. The shape-parameters of the lupus nephritis, acute post APGN and IgA nephropathy

Immune complex	shape-parameters (nm)		
	X-axis	Y-axis	Z axis
lupus nephritis	6.8	7.2	8.6
APGN	5.3	5.8	6.0
IgA nephropathy	5.7	6.4	6.6

Since the basic finding is the entrapment, the nanostructure of the glomerulus correlative in the germesis of the glomerulonephritis can be expected. Basically, if the immune complex is smaller than the size of glomerular pore, entrapment will not occur. Here, the author uses a computation technology to determine the size of the immune complex in important glomerulonephritis diseases.

It can be seen that the sizes of the immune complexes are larger than the glomerular pore size (about 4.5 – 5.0 nm in diameter) warranting the importance of the size of the immune complex in the germesis of glomerulonephritis.

It hereby concludes that the size of the immune complex is an important factor in the development of immune glomerulonephritis diseases. However, other factors should also be considered. Immune response is another important germic factor in immune glomerulitis. For example, the pediatric population is more susceptible to the nephritogenic antigen than adults, because the positive rate of anit-nephritogenic antibody in pediatrics is less than that of adults [55]. This can be another explanation for why APGN is more common in pediatrics.

C. Kernicterus in Preterm Infant: An Explanation Based on Nanomaterial Composition of Blood Brain Barrier

Neonatal jaundice is an important problem in medicine [56 – 57]. Around 80% of preterm infants develop hyperbilirubinaemia characterized by jaundice [58]. This condition is most commonly treated with phototherapy and/or exchange transfusion [58]. One of the most fatal unwanted types of neonatal jaundice is kernicterus [56 – 57]. Historically, the existence of yellow staining and damage to the brain caused by unconjugated bilirubin was first described by Hervieux in 1847 [59]. Kernicterus, the technical term used to describe the intense yellow staining in the basal ganglia of the brain, was first used by Schmorl in 1903 [59]. Kernicterus is a preventable brain injury caused by severe jaundice, yet it remains a threat today [60]. "Recent progress in understanding bilirubin binding and neurotoxicity resulting from unbound or liberated unconjugated bilirubin, how bilirubin impacts the central nervous system in vivo and in vitro, and the use of new clinical tools in neonates, for example magnetic resonance imaging revealing bilateral lesions in globus pallidus and subthalamus, and abnormal brainstem auditory evoked potentials with normal inner ear function, may lead to improved detection and prevention of neurologic dysfunction and damage from bilirubin" [61]. However, the exact mechanism for the described germesis of bilirubin toxicity has never been clarified.

Here, the author tries to explain this finding based on the nanomaterial composition of the blood brain barrier. First the author found out the pore sizes of the blood brain barrier in preterm and term neonate. According to the search, the pore sizes of the blood brain barrier in preterm and term neonate are 1.2 nm and 0.8 nm [62 – 63], respectively. Then the author further reviewed for the sizes of conjugated and unconjugated bilirubin. According to the search, the sizes of conjugated and unconjugated bilirubin are 2.2 nm and 1 nm respectively [64]. Basically, passing through a pore, the molecule must have a smaller size than the pore size of the membrane. Concerning the diffusion, at the controlled temperature and other environmental factors, conjugated bilirubin is unable to pass through the blood brain barrier pore of either preterm or term neonate.

Table 6. Useful and not useful aspects of nanomedicine

Useful	Not useful
Extremely bad conditions such as cancer and presently untreatable viral infections can be treated easily by modifying the body's genetic material. Disease elimination will become normal and easy. There is no necessity to be worried about living with poor health conditions. Nanomedicine can help early diagnosis of diseases before there are any symptoms. Nanomedicine can help administer medicines that are precisely targeted, called "magic bullet". Nanomedicine allows use of non-invasive imaging tools to demonstrate that the treatment was effective.	Probable toxicity of nanoparticles. The accidental unwanted modification consequences for the person or society. How does society ensures that the local government doesn't use inadequate much money to research methods that are not in the best interests of the citizens or are useless? There are also some ethical concerns on nanomedicine to be clarified.

Why is unconjugated bilirubin able to pass across the blood brain barrier of preterm but not term neonate? This can be a good explanation of why kernicterus is more common in preterm neonate and the cause is usually due to the unconjugated hyperbilirubinemia such as hemolysis.

NANOMEDICINE [65 – 66]

Nanomedicine is a new branch of medical science. Nanomedicine embraces five main sub-disciplines: a) analytical tools, b) nanoimaging, c) nanomaterials and nanodevices, d) new therapeutics and medicine delivery systems, and e) clinical, regulatory, and toxicological issues. It can be said that the potential of nano-technologies raises great hopes. The short term potential benefits include therapies for cancer, antiviral and antifungal agents, arteriosclerosis, diabetes, and chronic lung diseases; while the long term potential benefits include gene therapy and cell repair.

A. Nanomedicine for Diagnosis

Nanotechnologies can help physicians' diagnosis at single-cell and molecule levels, and some can be integrated into present molecular diagnostic methods, such as biochips [67]. Nanoparticles, such as gold nanoparticles, silver nanoparticles, and quantum dots are the most generalized used, but various other nanotechnological devices for manipulation at the nanoscale as nanobiosensors are also useful for potential clinical applied usages [67]. In conclusion, nanotechnologies will extend the limits of present molecular diagnostics and help in point-of-care diagnostics, integration of diagnostics with therapeutics, and development of personalized medicine [67]. Jain, mentioned for his major role of advances in biochip and microarray technologies in the development of personalized medicine [68]. Jain proposed that biochips such as GeneChip, CYP450, electrochemical biochips, protein biochips, microfluidic biochips and nanotechnology-based biochips were setting an important role in molecular diagnostics, and their applied usage in point-of-care diagnosis is expected to help the development of personalized medicine [68]. Although the potential diagnostic applied usages are not limited, the most important present applied usages are foreseen in the areas of biomarker discovery, cancer diagnosis, and detection of infectious germs [67]. Further details on nanodiagnosis can be seen in another chapter in this book.

B. Nanomedicine for Therapy [69 -71]

With the advent on nanopharmacology, new medicine development becomes easier than in previous days. In addition, the pharmaceutical process obtains a lot of usefulness from the new nanodelivery system. Besides medicine therapy, nanomedicine can be applied in surgery such as photodynamic therapy. Further details on nanotherapy can be seen in another chapter in this book.

C. Nanomedicine for Prophylaxis [72 - 73]

Nanobiotechnology can also be used for the prophylaxis of disease. Vaccine development can also make use of the nanomedicine way. The new science called nanovaccinology is the new hope for present vaccine research.

D. Nanodentistry [74]

Nanodentistry will make probable the maintenance of comprehensive oral health by involving the use of nanomaterials, biotechnology (including tissue engineering) and, ultimately, dental nanorobotics (nanomedicine). It is another interesting branch of medicine at present.

REFERENCES

[1] Kane RS, Stroock AD. Nanobiotechnology: protein-nanomaterial interactions. *Biotechnol. Prog.* 2007 Mar-Apr;23(2):316-9.

[2] Medvedeva NV, Ipatova OM, Ivanov IuD, Drozhzhin AI, Archakov AI. Nanobiotechnology and nanomedicine. *Biomed Khim.* 2006 Nov-Dec;52(6):529-46.

[3] Blattler T, Huwiler C, Ochsner M, Stadler B, Solak H, Voros J, Grandin HM Nanopatterns with biological functions. *J. Nanosci. Nanotechnol.* 2006 Aug;6(8):2237-64.

[4] Groneberg DA, Giersig M, Welte T, Pison U. Nanoparticle-based diagnosis and therapy. *Curr. Medicine Targets.* 2006 Jun;7(6):643-8.

[5] Jain KK. The role of nanobiotechnology in medicine discovery. *Medicine Discov. Today.* 2005 Nov 1;10(21):1435-42.

[6] Penn SG, He L, Natan MJ. Nanoparticles for bioanalysis. *Curr. Opin. Chem Biol.* 2003 Oct;7(5):609-15.
[7] Laval JM, Mazeran PE, Thomas D. Nanobiotechnology and its role in the development of new analytical devices. *Analyst.* 2000 Jan;125(1):29-33.
[8] Patel GM, Patel GC, Patel RB, Patel JK, Patel M. Nanorobot: a tool in nanomedicine. *J. Medicine Target.* 2006 Feb;14(2):63-7.
[9] Baba Y. Nanotechnology in medicine. *Nippon Rinsho.* 2006 Feb;64(2):189-98.
[10] Cao R, Villalonga R, Fragoso A. Towards nanomedicine with a supramolecular approach: a review. *IEE Proc Nanobiotechnol.* 2005 Oct;152(5):159-64.
[11] Patton JS. A historical perspective on convergence technology.*Nat Biotechnol.* 2006 Mar;24(3):280-1.
[12] Service RF. Chemistry. 'Grandfather of Nanotech' dies at 62.*Science.* 2005 Nov 4;310(5749):754.
[13] Lynch Z. Neurotechnology and society (2010-2060). *Ann N Y Acad Sci.* 2004 May;1013:229-33.
[14] Trgovcevic Z. 50 years of molecular biology. *Lijec Vjesn.* 1994 Nov-Dec;116(11-12):315-8.
[15] Fleischmann RD, Adams MD, White O, Clayton RA, Kirkness EF, Kerlavage AR, Bult CJ, Tomb JF, Dougherty BA, Merrick JM, et al Whole-genome random sequencing and assembly of Haemophilus influenzae Rd. *Science.* 1995 Jul 28;269(5223):496-512.
[16] Lewin R. National Academy looks at human genome project, sees progress.*Science.* 1987 Feb 13;235(4790):747-8.
[17] Johson RS. The human genome project: what impact on basic research? *FASEBJ.* 1987 Dec;1(6):502-5.
[18] Yang ZR, Hamer R. Bio-basis function neural networks in protein data mining. *Curr. Pharm. Des.* 2007;13(14):1403-13.
[19] Ahrens CH, Wagner U, Rehrauer HK, Turker C, Schlapbach R. Present challenges and approaches for the synergistic use of systems biology data in the scientific community. *EXS.* 2007;97:277-307.
[20] Jordan BR. 'Genomics': buzzword or reality? *J. Biomed. Sci.* 1999 May-Jun;6(3):145-50
[21] Esser A. The one-eyed Lycurgus and other deities of light. *Klin. Monatsblatter. Augenheilkd Augenarztl Fortbild.* 1954;125(3):361-3.
[22] Weyant RG. Lycurgus: the Father of Applied Psychology? *Am. Psychol.* 1967 Jun;22(6):432-4.
[23] Josef W. Grinding Way of the Diatreta Glasses. | [SCHLEIFTECHNIK DER DIATRETGLAESER]. Glastech Ber 1978; 51 (5): 130-136.
[24] Hornyak GL, Patrissi CJ, Oberhauser EB, Martin CR, Valmalette JC, Lemaire L, Dutta J, Hofmann H. Effective medium theory characterization of Au/Ag nanoalloy-porous alumina composites. *Nanostructured Mat.* 1997; 9 (1-8): 571-574.
[25] Ruff AJ. Breastmilk, breastfeeding, and transmission of viruses to the neonate. *Semin Perinatol.* 1994;18: 510-6.
[26] Lamounier JA, Moulin ZS, Xavier CC. Recommendations for breastfeeding during maternal infections. *J Pediatr* (Rio J). 2004;80(5 Suppl):S181-8World Health Organization.

[27] Oftedal OT. The mammary gland and its starting point during synapsid evolution. *J. Mammary Gland Biol. Neoplasia*. 2002;7:225-52
[28] Dupont C, Heyman M. Food protein-induced enterocolitis syndrome: laboratory perspectives. *J. Pediatr. Gastroenterol. Nutr.* 2000;30 Suppl:S50-7.
[29] Abraham P, John TJ. Hepatitis C : a review with particular reference to the Indian scenario. *Indian J. Med .Microbiol.* 1995; 13: 5 -1 4.
[30] Sakugawa H, Yamashiro T, Maeshiro T. Hepatitis D and Hepatitis E. *Nippon Naika Gakkai Zasshi.* 2004 Nov 10;93(11):2351-6
[31] Sagir A, Kirschberg O, Heintges T, Erhardt A, Haussinger D. SEN virus infection. *Rev. Med Virol.* 2004 May-Jun;14(3):141-8.
[32] Abe K. Infection of TT virus. *Nippon Rinsho*. 2003 Feb;61 Suppl 2:254-60.
[33] Pawlowska M, Halota W. CMV infections. *Przegl Epidemiol.* 2004;58 Suppl 1:17-21
[34] Filteau SM. Role of breast-feeding in managing malnutrition and infectious disease. *Proc. Nutr. Soc.* 2000;59:565-72.
[35] Fang F, Zhong W, Wang H. Study on infection of transfusion transmitted virus in serum, breast milk of pregnant women and umbilical venous blood from their newborns. Zhonghua Fu Chan Ke Za Zhi. 2001;36:330-2.
[36] Lawrence RM, Lawrence RA. Breast milk and infection. *Clin. Perinatol.* 2004 ;31:501-28
[37] Goldfarb J. Breastfeeding. AIDS and other infectious diseases. *Clin. Perinatol.* 1993;20:225-43.
[38] Stiehm ER, Keller MA. Breast milk transmission of viral disease. *Adv. Nutr. Res.* 2001;10:105-22.
[39] Ohka S, Matsuda N, Tohyama K, Oda T, Morikawa M, Kuge S, Nomoto A. Receptor (CD155)-dependent endocytosis of poliovirus and retrograde axonal transport of the endosome. *J. Virol.* 2004 Jul;78(13):7186-98.
[40] 6. Fenner F, Henderson DA, Arita I, Jezek Z, Ladnyi ID. Smallpox and its eradication. Geneva: World Health Organization, 1988.
[41] Yamada E, Montoya M, Schuettler CG, Hickling TP, Tarr AW, Vitelli A, Dubuisson J, Patel AH, Ball JK, Borrow P. Analysis of the binding of hepatitis C virus genotype 1a and 1b E2 glycoproteins to peripheral blood mononuclear cell subsets. *J. Gen. Virol.* 2005;86:2507-12.
[42] Barrera A, Guerra B, Notvall L, Lanford RE. Mapping of the hepatitis B virus pre-S1 domain involved in receptor recognition. *J. Virol.* 2005;79:9786-98.
[43] Omerovic J, Lev L, Longnecker R. The amino terminus of Epstein-Barr virus glycoprotein gH is important for fusion with epithelial and B cells. *J. Virol.* 2005;79:12408-15.
[44] Lopper M, Compton T. Coiled-coil domains in glycoproteins B and H are involved in human cytomegalovirus membrane fusion. *J. Virol.* 2004;78:8333-41.
[45] Stagno S, Reynolds DW, Pass RF, Alford CA. Breast milk and the risk of cytomegalovirus infection. *N. Engl. J. Med.* 1980;302:1073-1076.
[46] Alford CA, Britt WJ. Cytomegalovirus. In: Fields BN, Knipe DM, Chanock RM, Hirsh MS, Melnich JL, Monath TP, Roizman B, eds. *Fields Virology*. 2nd ed. New York: Raven Press, 1990; 1981-2010.

[47] Minamishishima I, Ueda K, Minematsu T, Minamishishima Y, Unemoto M, Take H, Kuraya K. Role of breast milk in acquisition of cytomegalovirus infection. *Micobiol. Immunol.* 1994;38:549-552.

[48] Gambarotto K, Ranger-Rogez S, Aubard Y, Piver P, Duffetelle B, Delpeyroux C, Roussanne MC, Nicot T, Denis F. Primary cytomegalovirus infection and pregnant women: epidemiological study on 1100 women at Limoges. *Pathol. Biol.* 1997;45:453-461.

[49] Schroter M, Polywka S, Zollner B, Schafer P, Laufs R, Feucht HH. Detection of TT virus DNA and GB virus type C/Hepatitis G virus RNA in serum and breast milk: determination of mother-to-child transmission. *J. Clin. Microbiol.* 2000;38:745-7.

[50] Rosengard AM, Liu Y, Nie Z, Jimenez R. Variola virus immune evasion design: expression of a highly efficient inhibitor of human complement. *Proc. Natl. Acad. Sci. U S A* 2002;99:8808-13.

[51] Nangaku M, Couser WG. Mechanisms of immune-deposit formation and the mediation of immune renal injury. *Clin. Exp. Nephrol.* 2005;9:183-91.

[52] Turnberg D, Cook HT. Complement and glomerulonephritis: new insights. *Curr. Opin. Nephrol .Hypertens.* 2005 ;14:223-8.

[53] Berman HM, Westbrook J, Feng Z, Gilliland G, Bhat TN, Weissig H, Shindyalov IN, Bourne PE. The Protein Data Bank. *Nucleic Acids Research.* 2000; 28: 235-242.

[54] Rougier JP, Ronco P. Germesis of immune glomerulonephritis. *Rev. Prat.* 2003;53:1998-2004.

[55] Yoshizawa N, Yamakami K, Fujino M, Oda T, Tamura K, Matsumoto K, Sugisaki T, Boyle MD. Nephritis-associated plasmin receptor and acute poststreptococcal glomerulonephritis: characterization of the antigen and associated immune response. *J. Am. Soc. Nephrol.* 2004;15:1785-93.

[56] Wiwanitkit V. Neonatal Jaundice. 46th Annual conference of the Indian Society of Haematology and Transfusion medicine. Imphal, Manipur on 14th to 16th January 2006

[57] Shapiro SM, Bhutani VK, Johnson L. Hyperbilirubinemia and kernicterus. *Clin. Perinatol.* 2006; 33:387-410.

[58] Truman P. 2006. Jaundice in the preterm infant. *Paediatr. Nurs.*, 18:20-2.

[59] Juretschke LJ. 2005. Kernicterus: still a concern. *Neonatal Netw*, 24:7-19.

[60] Shapiro SM. 2003. Bilirubin toxicity in the producing nervous system. *Pediatr Neurol*, 29:410-21.

[61] van Bree JB, de Boer AG, Danhof M, Ginsel LA, Breimer DD. Characterization of an "in vitro" blood-brain barrier: effects of molecular size and lipophilicity on cerebrovascular endothelial transport rates of medicines. *J. Pharmacol. Exp. Ther.* 1988; 247:1233-9.

[62] Calhoun DA, Murthy SN, Bryant BG, Luedtke SA, Bhatt-Mehta V. Characterization of an "in vitro" blood-brain barrier: effects of molecular size and lipophilicity on cerebrovascular endothelial transport rates of medicines. *J. Pharmacol. Exp. Ther.* 2006; 247:1233-9.

[63] Seyfried H, Klicpera M, Leithner C, Penner E. 1976. Bilirubin metabolism. *Wien Klin Wochenschr.* 88:477-82.

[64] Murthy SK. Nanoparticles in novel medicine: state of the art and future challenges. *Int J. Nanomedicine.* 2007;2(2):129-41.

[65] Pauwels EK, Erba P Towards the use of nanoparticles in cancer therapy and imaging. *Medicine News Perspect.* 2007 May;20(4):213-20.
[66] Jain KK. Applied usages of Nanobiotechnology in Clinical Diagnostics.*Clin. Chem.* 2007 Sep 21;
[67] Jain KK. Applied usages of biochips: from diagnostics to personalized medicine. *Curr. Opin. Medicine Discov. Devel.* 2004 May;7(3):285-9.
[68] Moghimi SM, Hunter AC, Murray JC. Nanomedicine: present status and future prospects. *FASEB J.* 2005 Mar;19(3):311-30.
[69] Emerich DF. Nanomedicine--prospective therapeutic and diagnostic applied usages. *Expert Opin. Biol. Ther.* 2005 Jan;5(1):1-5.
[70] Nakayama M, Okano T. Medicine delivery systems using nano-sized medicine carriers. *Gan To Kagaku Ryoho.* 2005 Jul;32(7):935-40.
[71] Lee LA, Wang Q. Adaptations of nanoscale viruses and other protein cages for medical applied usages. *Nanomedicine.* 2006 Sep;2(3):137-49.
[72] 73. Lori F, Calarota SA, Lisziewicz J. Nanochemistry-based immunotherapy for HIV-1.*Curr. Med. Chem.* 2007;14(18):1911-9.
[73] 74. Freitas RA Jr. Nanodentistry. *J. Am. Dent Assoc.* 2000 Nov;131(11):1559-65.

Chapter 2

NANOSTRUCTURE MATERIALS AND NANOMODIFICATION TECHNIQUES

COMMON NANOSTRUCTURE MATERIALS

With the advent of nanotechnology, a number of nanostructure materials have been produced. The examples of commonly used nanomaterials will be discussed here in detail.

1. Gold Nanoparticle

Gold nanoparticle is one of the most used nanomaterials at present [1 - 2]. Due to the very small size of the gold nanoparticle, the surface area significantly increases compared to general gold. Therefore, it can be used as a sensor. Of interest, the different sizes of gold nanoparticles result in a difference in color. The change of the color of gold nanoparticles from pink at a very small size to gray at a larger size can be seen. This phenomenon can be used as an indicator for a naked eye detection system. In addition to the naked eye detection system, gold nanoparticles can be used for detection purposes. The DNA linker system is another interesting applied usage. Based on the previously described principle, gold nanoparticle with a DNA linker can result in a large size gold nanoparticle complex. The change in color can be seen. A change from pink color to gray color can be a basic observation by the naked eye system (Figure 1). Gold nanoparticle is also applied as an enhancing apparatus for electrocatalysis and electrochemical sensors [1 - 2]. In addition, Han et al also presented that functionalized gold nanoparticles represented highly attractive and useful candidates in the applied usages of medicine delivery owing to their unique dimensions, tunable functionalities on the surface, and controllable medicine release [2].

Table 1. Some interesting reports on the gold nanoparticle and DNA linker system

Authors	Details
Qin and Yung [3]	This study reported an interesting new approach to generate gold nanoparticle-DNA conjugates bearing specially designed DNA linker molecules that can be used as building blocks to construct nanoassemblies with precisely controlled structure or as nanoprobes for quantitative DNA sequence detection analysis [3]. In this work, gold nanoparticle-DNA conjugates bearing a specific number of long double-stranded DNA strands were treated by gel electrophoresis, and a restriction endonuclease enzyme was then used to manipulate the length of the nanoparticle-bound DNA [3]. Qin and Yung reported that this enzymatic cleavage was warranted by gel electrophoresis, and digestion efficiency of 90% or more was getd [3].
Wang et al. [4]	In this work, gold nanoparticles were attached to a microarray via base pairing between complementary DNA sequences on the array and on the particles [4]. Wang et al found that these particles could be detected by light scattering measurements following an enhancement step, in which silver islands were deposited on top of the gold particles [4].
Becker et al. [5]	Functional immobilization of the small GTPase Rab6A on DNA-gold nanoparticles by using a site-specifically attached poly(ethylene glycol) linker and thiol place-exchange reaction was reported by Becker et al in this article [5].
Zhou and Zhou [6]	Zhou and Zhou developed a new class of silica core-shell nanoparticles that encapsulated with cyanine dyes and applied the dye-doped nanoparticles as tagging in the DNA microarray-based bioanalysis [6]. In this work, Zhou and Zhou demonstrated developed nanoparticles posing core-shell structure consisting of 15-nm gold colloidal cores with 95 dye-alkanethiol (dT)20 oligomers chemisorbed on each gold particle surface, and 10-15-nm silica coatings bearing thiol functional groups [6]. This work offers useful prospects for applying dye-doped nanoparticles as tagging for gene profiling based on DNA microarray technology [6].
Harris and Kiang [7]	Harris and Kiang reported experimental observations on the effect of disorder on the phase property of DNA-linked nanoparticle assemblies [7]. Harris and Kiang observed an unusual potential of a nonmonotonic "zigzag" pattern in the melting temperature as a function of DNA linker length [7]. Harris and Kiang said that linker DNA resulting in unequal DNA duplex lengths introduced disorder and lowered the melting temperature of the nanoparticle system [7].
Pan et al. [8]	In this study, DNAs were directly attached to gold surfaces through S-gold bonds by modifying 5'-end of DNA with a mercapto group [8]. Pan et al found that the pH of the reaction mixture played an important role in DNA-gold conjugate fabrication and the DNA-Au conjugates were completely obtained in pure forms and, with the use of this new type conjugate, gold-DNA-gold nanostructure was treated [8].
Jung et al. [9]	In this work, a biolinker for efficient antibody immobilization was treated by site-specific coupling of protein G to DNA oligonucleotide [9]. Jung et al reported that antibody targeting on glass slides could be getd by using this linker system without modifying or spotting antibodies, and the protein G-DNA conjugate brought a simple but effective method to label DNA-functionalized gold nanoparticles with target antibodies [9]. Jung et al concluded that the DNA-linked protein G construct introduced in this study offered a useful method to manage antibody immobilization in many immunoassay systems [9].

Authors	Details
Qin and Yung [10]	Qin and Yung report a gold nanoparticle-based method that allows DNA detection and quantification, and is capable of single nucleotide polymorphism (SNP) discrimination [10]. According to this work, for SNP detection, single base mismatch discrimination was getd for both the end- and center-base mismatch by Qin and Yung [10]. Qin and Yung summarized that the method might be useful for the development of a simple and quantitative DNA detection assay [10].

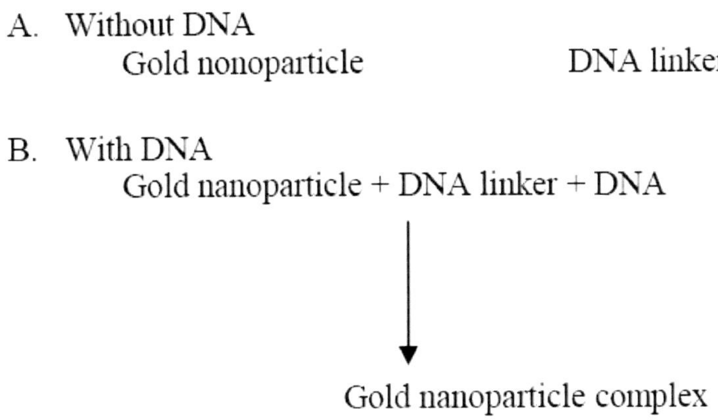

Figure 1. Diagram showing gold nanoparticle and DNA linker interaction.

In medical diagnostics, gold nanoparticle can be a useful tool for laboratory analysis. A good example is the applied usage in the diagnosis of proteinuria. Wiwanitkit et al. also recently reported this in the development of a gold nanoparticle amplified agglutination system for weak blood group determination [11]. Nevertheless, it should also be noted that gold nanoparticle is not actually a new thing. Indeed, our ancestors have used gold nanoparticle for a long time. The Lycurgus [12] cup (A.D.300) is the best evidence. In addition, gold nanoparticle is also commonly used as an important ingredient in many classical Chinese medicine remedies.

2. Silver Nanoparticle

Silver nanoparticle is another commonly used nanomaterial [13 – 14]. The different sizes of silver nanoparticles also results in different colors. It should be noted that silver nanoparticle, which has a size smaller than 10 nm has a golden yellow color while the color changes into red and black in larger sizes. Chen et al. studied the morphology effects on the optical properties of silver nanoparticles [15]. According to this work, solutions of the silver particles revealed distinct surface plasmon resonant absorption signatures that were directly correlated with the morphological symmetries [15]. In addition, nonlinear optical property for suspensions of these nanostructures, for nanosecond pulses at 532 and 1064 nm, were also correlated to plasmon resonances detected by shape and size [15]. There was also another interesting study by Whelan et al. [16]. According to this work, polyvinylalcohol -stabilized silver nanoparticles were treated by a seeding method [16]. Whelan et al. reported that the samples revealed two bands in the visible absorption spectrum: one in the 410-440-nm region and a second peak between 500 and 600 nm [16]. Recently, Shankar et al reported on the use of Geranium (*Pelargonium graveolens*) leaf extract in the extracellular synthesis of silver nanoparticles [17]. This study represents an important advance in the use of plants over microorganisms in the biosynthesis of metal nanoparticles [17]. Ankamwar et al reported the extracellular synthesis of silver nanoparticles using Emblica *Officinalis* (amla, Indian Gooseberry) fruit extract as the reducing agent to synthesize nanoparticles, their subsequent phase transfer to an organic solution and the transmetallation reaction of hydrophobized silver nanoparticles with hydrophobized chloroaurate ions [18]. Ankamwar et al said that transmetallation reaction between hydrophobized silver nanoparticles and hydrophobized chloroaurate ions in chloroform caused the formation of gold nanoparticles [18].

An interesting aspect of silver nanoparticle is the fact that the increased surface area of nanoparticle significantly increases the catalytic activity, therefore silver nanoparticle is commonly used as the magic catalyst. It is proven that silver nanoparticle is an effective bactericidal agent. Panacek et al reported that antibacterial activity of silver nanoparticles was found to be dependent on the size of silver particles [19]. Panacek et al. said that a very low concentration of silver gave antibacterial characteristic [19]. Gogoi et al indicated that silver nanopaticles, above a certain concentration, not only had bactericidal acitivity but also were found to reduce the sizes of treated bacteria in comparison to untreated ones [20]. Indeed, silver is an important composition of many anti-infective dyes such as silver sulfadiazene. The mechanism of the bactericidal process is believed to be an attachment on the bacterial cell membrane due to the generation of free radicals. Pal et al. investigated the antibacterial properties of differently shaped silver nanoparticles against the gram-negative bacterium *Escherichia coli*, both in liquid systems and on agar plates to answer the query if antibacterial activity of silver nanoparticles depended on the shape of the nanoparticle [21]. According to this work, Pal et al. demonstrated that silver nanoparticles underwent a shape-dependent interaction with the gram-negative organism *E. coli* [21]. In another quoeted work, Sharverdi et al. also said that the antibacterial activities of penicillin G, amoxicillin, erythromycin, clindamycin, and vancomycin were increased in the existence of silver nanoparticle [22].

At present, silver nanoparticle is commonly used in the textile industry. Nanocloth is a new applied usage. For example, nanoshirt and nanosock are mentioned for their antibacterial properties. The prevention of unwanted odor due to bacterial fermentation in sweat and dirt contaminations can be derived and these clothes require infrequent washing. In addition to nanocloth, there are other silver nanoparticle products at present including silver nanoparticle spray for contol of bacterial contamination, silver nanoparticle-coated washing machine, and a silver nanoparticle-coated refrigerator. In 2006, Cohen et al. analyzed a polypropylene mesh coated with nanosilver particles as a means of preventing such infections [23]. Cohen et al. found that the silver-coated mesh demonstrated significant bactericidal activity compared with uncoated polypropylene mesh [23]. In addition, Cohen et al. also detected that nanocrystalline silver particles might decrease the incidence of postoperative prosthetic mesh infections and be useful as a coating for other prosthetic materials [23].

Similar to gold nanoparticle, silver nanoparticle can also be used in the DNA linker system. Some interesting reports on silver nanoparticle and the DNA linker system are presented in Table 2.

Table 2. Some interesting reports on the silver nanoparticle and DNA linker system

Authors	Details
Lee et al. [24]	Lee et al reported a new method for preparing silver nanoparticle-oligonucleotide conjugates that were based upon DNA with cyclic disulfide-anchoring groups [24]. Lee et al mentioned that this work was an important step toward using silver nanoparticle-oligonucleotide conjugates for a multiple purposes, including molecular diagnostic labels, synthons in programmable materials synthesis approaches, and functional components for nanoelectronic and plasmonic devices [24].
Shang et al. [25]	Shang et al reported their findings in the preparation of DNA-silver nanohybrids in multilayer nanoreactors by in electrochemical reduction, characterization, and applied usage [25].
Sabanayagam and Lakowicz [26]	Sabanayagam and Lakowicz examined metal-amplified fluorescence by spotting biotinylated, singly-labeled 23 bp DNAs onto avidin-coated SIF substrates [26]. In this work, increasing the sensitivity of DNA microarrays by metal-amplified fluorescence using surface-bound silver nanoparticles was noted [26].
Dawn and Nandi [27]	In this work, a new nano-biocomposite of silver and poly(o-methoxy aniline) (POMA)/DNA hybrid was treated by adding DNA solution to an aqueous solution of POMA (emeraldine base, EB) and AgNO(3) mixture. Then the mixture was stable for 10 days and was freeze-dried to form the hybrid nanocomposite [27]. Dawn and Nandi reported that at a higher applied voltage, semiconducting property characterizing the large band gap semiconducting property of the nano-biocomposite could be observed [27].

3. Iron Nanoparticle

Iron, the most ubiquitous of the transition metals and the fourth most plentiful metal in the Earth's crust, is the structural backbone of our novel infrastructure [28]. Iron nanoparticle is another important metallic nanoparticle. Finely divided iron has long been known to be pyrophoric, which is a major reason that iron nanoparticles have not been more completely studied to date [28]. The syntheses of amorphous Fe(2)O(3) nanoparticles of varying size and morphology, their magnetic properties, crystallization mechanism, and applied usages are a present focus of nanobiotechnology [29]. Superparamagnetic iron oxide particles have multiple applied usages in molecular and cellular imaging [30]. Thorek et al. noted that superparamagnetic iron oxide nanoparticles demonstrated their utility as an important tool for enhancing magnetic resonance contrast, allowing researchers to monitor not only anatomical changes, but physiological and molecular changes too [31]. At present, superparamagnetic iron oxide particles are commonly used. This is because of the nontoxicity and biocompatibility demand that mainly iron oxide-based materials are predominantly used, despite some attempts to develop more magnetic nanomaterials based on cobalt, nickel, gadolinium and other compounds [32]. For all these applied usages, the material used for surface coating of the magnetic particles must not only be nontoxic and biocompatible but also allow a targetable delivery with particle localization in a specific area [32].

Table 3. Some reports on applied usage of iron nanoparticle in nanomedicine

Authors	Details
Hu et al. [33]	In this work, magnetic sponge-like hydrogels (ferrosponges) were fabricated by using an in-situ synthesis of magnetic iron nanoparticles in the existence of various concentrations of gelatin [33]. Hu et al. noted that a new medicine delivery system could be designed for medical uses by using these peculiar magnetic sensitive properties of the ferrosponges [33].
Ge et al. [34]	In this work, one-step synthesis of highly hydro-soluble magnetite colloidal nanocrystals was reported [34]. Ge et al. said that the monodisperse, highly hydro-soluble, superparamagnetic, and biocompatible magnetite nanocrystals should find sudden important biomedical applied usages [34].
Arias et al. [35]	A reproducible method for the preparation of mixed colloidal nanoparticles, consisting of of a magnetic carbonyl iron nucleus and a biocompatible ethylcellulose latex shell, was described in this article [35]. Arias et al. said that this carbonyl iron/ethylcellulose core/shell nanoparticle was useful for biomedical applied usages [35].
Ito et al. [36]	Ito et al. reported magnetite nanoparticle-loaded anti-HER2 immunoliposomes for combination of antibody therapy with hyperthermia [36]. Ito et al. suggested that this new therapeutic tool was applied to treatment of HER2-overexpressing cancer [36].

Table 4. The details and the required energy for complex formation in each step of formation of complex

Items	Detials
Bond breaking*	O – O, e- from Fe (II)
Bond forming**	-
Obtaining energy	172.5 kCal/mol + 1 eV/mol
Giving out energy	0
Required energy***	172.5 kCal/mol + 1 eV/mol

*bond breaking obtaining energy
**bond forming giving out energy
***required energy = obtaining energy - giving out energy

4. Titanium Oxide Nanoparticle

Titanium dioxide (TiO2) is a useful material for many emerging applied usages [37]. Nanotitenium oxide (NanoTiO2) is a commonly used particle in nanocloth [37]. NanoTiO2 is mentioned for its good antibacterial activity. An effective organic degradation process is documented [38]. Similar to silver nanoparticle, the antibacterial property of nanotitenium oxide is proposed. To help the reader better understand the mechanism of the bactericidal effect of titanium oxide nanoparticle, the author will further explain the quantum mechanics for what occurs in the mentioned reaction. Concerning the organic degradation reaction, photocatalytic activity occurs [39]. After photostimulation, the transfer of electrons (e-) from nano-particles brings a reduction of oxygen composition to produce hydrogen peroxide (H2O2) which, in the existence of iron ions (Fe (II)) in bacterial cell membrane, can then form hydroxyl radicals (OH-) via the Fenton reaction. This process is called the photolytic-

Fenton process. The quantum chemical analysis for total reaction was performed according to the classical bonding theory. The required energy for reaction was calculated. The details and the required energy for the reaction are presented in Table 4. The required energy for complex formation is 172.5 kCal/mol + 1 eV/mol.

Classification of Nanostructure Materials [40]

There are many systems for classification of nanostructure materials. The two most generalized used systems are dimension of material-based system and characteristic of material-based system. Concerning dimension of material-based system, there are three groups of nanostructure materials. The first group is one-nano dimension. A good example is nanofilm. The second group is two-nano dimensions. This group has a common tubular fiber structure. Examples of nanostructure materials in this group are nanofilter, nanowire and carbon nanotube. The third group is three-nano dimensions. Examples are nanoparticle, nanopowder, nanocapsule, quantum dot, nanopore, nanostructure, dendrimer and fullerene.

Table 5. Examples of nanostructure materials classified by characteristic of material–based system

Groups	Examples
organic nanoparticle	dendrimer, polymer
inorganic nanoparticle	gold nanoparticle, silver nanoparticle, iron oxide
organic / inorganic hybrid	nanocomposite, core-shell type
carbon – based nanoparticle	functionalized fullerene
liposome	functionalized liposome, inclusion complex
biological nanoparticle	protein and peptide - based nanoparticles with other biological components

Concerning the characteristics of the material-based system, six main groups are organic nanoparticle, inorganic nanoparticle, organic/inorganic hybrids, carbon-based nanomaterial, lyzosome and biological nanoparticles. Examples of nanostructure materials in each group are presented in Table 5.

Preparation of Nanostructure Materials [40 - 42]

The details for preparation of different nanostructure materials are different. However, a simple common way can be used. There are two main types of preparations, preparation form natural materials and new synthesis. For preparation of nanostructure materials from natural bulk material, the first step is desizing the bulk material. A homogenizer can be used. The stabilization of the treated nanomaterial is of concern. Different buffers can impact the property of the treated material and this has a further consequence on the usage. For example, preparation of gold nanoparticle in a citrate buffer solution can induce hemolysis of red blood cell if applied into blood sample. Another point is that the stability of the treated materials must be tested. It is necessary to warrant that the treated material does not aggregate to form a large size material with a loss of its nanomaterial property. The final step is the warrantation of the size of the treated material. The sizing machine can help this step. The size measurement can warrant that the desired nanolevel is reached and it can bring basic information for following up the stability of the treated nanomaterial as previously described.

Considering new synthesis, the polymerization way is the basic requirement. New synthesized nanomaterials are usually designed for a specific purpose. The design is the first step because this step is the determination for the further synthetic procedure. The design should focus on these factors: baseline input materials, processing and designed finalized products. The new synthesis is usually used for polymerized nanomaterial as previously mentioned. The nanonodification way should be applied in design. The details will be further discussed in the next heading. The next step in designing is polymerization and shaping. This step is important and it can change the resulting material into a different dimensional appearance. However, most of new syntheses result in two or three dimensional nanoparticles. The most generalized synthesized nanomaterials are in the spherical shape. Shelling, covering a nanomaterial with another nanomaterial (as egg shell), is generalized used. The test for stability and sizing is also required, similar to the preparation from natural materials. There is an additional concern of new synthesis. It should be noted that although the new synthesis is ful and the resulting product is fit as designed, the new synthesized material might not be feasible for real usage. The concept of size, which was mentioned in the first chapter of this book, must be strictly followed. As a case study, the author was consulted from a scientist on the problem of polyethylene core shell for gene therapy [43 - 44]. The problem is that the synthesized material cannot transfer the gene into the cell. The main problem is that the polyethylene core shell has a larger diameter than the cell membrane pore diameter, therefore the polyethylene core shell is not able to pass into the cell.

Nanomodification Techniques [45]

There are many nanomodification techniques at present. However, the three main important ones are nanoencapsulation, nanodevicing and nanocoating.

A. Nanoencapsulation

Polymeric nanoparticles have been extensively studied as particulate carriers in the pharmaceutical and medical fields, because they show promise as a medicine delivery systems as a result of their controlled- and sustained-release properties, subcellular size, and biocompatibility with tissue and cells [46 - 47]. Dispersed polymer nanocapsules can serve as nano-sized medicine carriers to get controlled release too as efficient medicine targeting [48]. The dispersion stability and the primary physiological response are mainly detected by the type of surfactant and the nature of the outer coating [48]. Nanoencapsulation is used for synthesis of a new nanomaterial from a combination of two or more baseline input materials. Shelling is the basic procedure. The functionalized liposome is the best example. This nanoencapsulation way is useful for development of a medicine and active gradient transportation system [47 – 48]. This way is commonly used for producing new anticancer medicines [50]. Of interest, at temperatures below the gel-liquid crystalline transition, T(m), the interdigitated lipid-ethanol sheets are rigid and flat; when the temperature is increased above T(m), the sheets become flexible and close on themselves and the surrounding solution to form closed compartments [50]. During this closure, the sheets can entrap other vesicles, biological macromolecules, or colloidal particles resulting in a special nanoencapsulation namely vesosome [50]. The vesosome structure has many uses including steric stabilization, pH loading of medicines, and intrinsic biocompatibility [50].

B. Nanodevicing [51]

Nanodevicing is the modification of nanostructure material into a nanodevice. A combination of basic knowledge in nanobiotechnology, nanoelectronics and nanomaterial science is required for nanodevicing. Generalized used examples of nanodevicing are thin film sensor production from nanofilm, solar cell production and electrical transducer polymer production.

C. Nanocoating [51]

Nanocoating is a commonly used nanomodification technique. It is similar to, but not the same as, nanoencapsulation. Nanocoating does not result in a shelled material. It is not necessary that the resulting product be a spherical shape material. In addition, the resulting finalized material is not required to be a nanomaterial. A good example is nanofiber coating with silver nanoparticle in textile production to result in nanocloth to be worn. There are many recent interesting reports on nanocoating. In 2004, Beck et al. designed and treated new nanoparticle-coated medicine-loaded inorganic microparticles by spray-drying using diclofenac as a medicine model [52]. Beck et al. concluded that the use of diclofenac as a hydrophobic model conducted to powders presenting good gastroresistance when the nanocapsules and triacetin were employed [52]. Monteiro et al. studied zinc sulfide nanocoating of silica submicron shperical objects using a single-source method [53]. Monteiro et al said that the silica surfaces became homogeneously coated with zinc sulfide as the reaction time was increased and this coating consisted of clusters of small ZnS particles whose dimensions were within the nanometric range [53].

References

[1] Guo S, Wang E. Synthesis and electrochemical applied usages of gold nanoparticles. *Anal. Chim. Acta.* 2007 Aug 29;598(2):181-92.

[2] Han G, Ghosh P, Rotello VM. Functionalized gold nanoparticles for medicine delivery. *Nanomed.* 2007 Feb;2(1):113-23.

[3] Qin WJ, Yung LY. Nanoparticle-DNA conjugates bearing a specific number of short DNA strands by enzymatic manipulation of nanoparticle-bound DNA. *Langmuir.* 2005 Nov 22;21(24):11330-4.

[4] Wang Z, Lee J, Cossins A, Brust M. Selective enzymatic cleavage of gold nanoparticle-labelled DNA on a microarray. IEE Proc *Nanobiotechnol.* 2005 Apr;152(2):85-8.

[5] Becker CF, Marsac Y, Hazarika P, Moser J, Goody RS, Niemeyer CM. Functional immobilization of the small GTPase Rab6A on DNA-Gold nanoparticles by using a site-specifically attached poly(ethylene glycol) linker and thiol place-exchange reaction. *Chembiochem.* 2007 Jan 2;8(1):32-6.

[6] Zhou X, Zhou J. Improving the signal sensitivity and photostability of DNA hybridizations on microarrays by using dye-doped core-shell silica nanoparticles. *Anal.Chem.* 2004 Sep 15;76(18):5302-12.

[7] Harris NC, Kiang CH. Disorder in DNA-linked gold nanoparticle assemblies.*Phys. Rev.* Lett. 2005 Jul 22;95(4):046101.

[8] Pan F, Sumaoka J, Hirano T, Okutsu Y, Takeuchi O, Shigekawa H, omiyama M. Fabrication of Au-DNA-Au nanostructure with new-type DNA-Au conjugate. *Nucleic Acids Symp. Ser* .(Oxf). 2006;(50):317-8.

[9] Jung Y, Lee JM, Jung H, Chung BH. Self-directed and self-oriented immobilization of antibody by protein G-DNA conjugate. *Anal .Chem.* 2007 Sep 1;79(17):6534-41.

[10] Qin WJ, Yung LY. Nanoparticle-based detection and quantification of DNA with single nucleotide polymorphism (SNP) discrimination selectivity. *Nucleic Acids Res.* 2007;35(17):e111.

[11] Wiwanitkit V, Seereemaspun A, Rojanathanes R. Increasing the gglutination reaction in slide test for weak B blood group by gold nanoparticle solution: The first world report. *J. Immunol. Methods.* 2007 Oct 4;

[12] Esser A. The one-eyed Lycurgus and other deities of light. *Klin Monatsblatter Augenheilkd Augenarztl Fortbild.* 1954;125(3):361-3.

[13] Evanoff DD Jr, Chumanov G. Synthesis and optical properties of silver nanoparticles and arrays. *Chemphyschem.* 2005 Jul 11;6(7):1221-31.

[14] Lee KS, El-Sayed MA. Gold and silver nanoparticles in sensing and imaging: sensitivity of plasmon response to size, shape, and metal composition. *J. Phys. Chem. B.* 2006 Oct 5;110(39):19220-5.

[15] Chen S, Webster S, Czerw R, Xu J, Carroll DL. Morphology effects on the optical properties of silver nanoparticles. *J. Nanosci. Nanotechnol.* 2004 Mar;4(3):254-9.

[16] Whelan AM, Brennan ME, Blau WJ, Kelly JM. Amplified third-order optical nonlinearity of silver nanoparticles with a tunable surface plasmon resonance. *J. Nanosci .Nanotechnol.* 2004 Jan-Feb;4(1-2):66-8.

[17] Shankar SS, Ahmad A, Sastry M. Geranium leaf assisted biosynthesis of silver nanoparticles. *Biotechnol. Prog.* 2003 Nov-Dec;19(6):1627-31.

[18] Ankamwar B, Damle C, Ahmad A, Sastry M. Biosynthesis of gold and silver nanoparticles using Emblica Officinalis fruit extract, their phase transfer and transmetallation in an organic solution. *J. Nanosci. Nanotechnol.* 2005 Oct;5(10):1665-71.

[19] Panacek A, Kvitek L, Prucek R, Kolar M, Vecerova R, Pizurova N, Sharma VK, Nevecna T, Zboril R. Silver colloid nanoparticles: synthesis, characterization, and their antibacterial activity. *J. Phys. Chem. B.* 2006 Aug 24;110(33):16248-53.

[20] Gogoi SK, Gopinath P, Paul A, Ramesh A, Ghosh SS, Chattopadhyay A. Green fluorescent protein-expressing Escherichia coli as a model system for investigating the antimicrobial activities of silver nanoparticles.*Langmuir.* 2006 Oct 24;22(22):9322-8.

[21] Pal S, Tak YK, Song JM. Does the antibacterial activity of silver nanoparticles depend on the shape of the nanoparticle? A study of the Gram-negative bacterium Escherichia coli. *Appl. Environ. Microbiol.* 2007 Mar;73(6):1712-20.

[22] Shahverdi AR, Fakhimi A, Shahverdi HR, Minaian S. Synthesis and effect of silver nanoparticles on the antibacterial activity of different antibiotics against Staphylococcus aureus and Escherichia coli. *Nanomedicine.* 2007 Jun;3(2):168-71.

[23] Cohen MS, Stern JM, Vanni AJ, Kelley RS, Baumgart E, Field D, Libertino JA, Summerhayes IC. In vitro analysis of a nanocrystalline silver-coated surgical mesh. *Surg. Infect (Larchmt).* 2007 Jun;8(3):397-403.

[24] Lee JS, Lytton-Jean AK, Hurst SJ, Mirkin CA. Silver nanoparticle-oligonucleotide conjugates based on DNA with triple cyclic disulfide moieties. *Nano. Lett.* 2007 Jul;7(7):2112-5.

[25] Shang L, Wang Y, Huang L, Dong S. Preparation of DNA-silver nanohybrids in multilayer nanoreactors by in situ electrochemical reduction, characterization, and applied usage. *Langmuir.* 2007 Jul 3;23(14):7738-44.

[26] Sabanayagam CR, Lakowicz JR. Increasing the sensitivity of DNA microarrays by metal-amplified fluorescence using surface-bound silver nanoparticles. *Nucleic Acids Res.* 2007;35(2):e13.

[27] Dawn A, Nandi AK. Formation of silver nanoparticles in deoxyribonucleic acid-poly(o-methoxyaniline) hybrid: a new nano-biocomposite. *J. Phys. Chem. B.* 2006 Sep 21;110(37):18291-8.

[28] Huber DL. Synthesis, properties, and applied usages of iron nanoparticles. *Small.* 2005 May;1(5):482-501.

[29] Machala L, Zboril R, Gedanken A. Amorphous iron(III) oxide--a review. *J. Phys. Chem.* B. 2007 Apr 26;111(16):4003-18.

[30] Corot C, Robert P, Idee JM, Port M. Recent advances in iron oxide nanocrystal technology for medical imaging. *Adv. Medicine. Deliv. Rev.* 2006 Dec 1;58(14):1471-504.

[31] Thorek DL, Chen AK, Czupryna J, Tsourkas A. Superparamagnetic iron oxide nanoparticle probes for molecular imaging. *Ann. Biomed. Eng.* 2006 Jan;34(1):23-38.

[32] Gupta AK, Naregalkar RR, Vaidya VD, Gupta M. Recent advances on surface engineering of magnetic iron oxide nanoparticles and their biomedical applied usages. *Nanomed.* 2007 Feb;2(1):23-39.

[33] Hu SH, Liu TY, Liu DM, Chen SY. Nano-ferrosponges for controlled medicine release. *J. Control Release.* 2007 Aug 28;121(3):181-9.

[34] Ge J, Hu Y, Biasini M, Dong C, Guo J, Beyermann WP, Yin Y.One-step synthesis of highly hydro-soluble magnetite colloidal nanocrystals. *Chemistry.* 2007;13(25):7153-61.

[35] Arias JL, Lopez-Viota M, Ruiz MA, Lopez-Viota J, Delgado AV. Development of carbonyl iron/ethylcellulose core/shell nanoparticles for biomedical applied usages. *Int. J. Pharm.* 2007 Jul 18;339(1-2):237-45.

[36] Ito A, Kuga Y, Honda H, Kikkawa H, Horiuchi A, Watanabe Y, Kobayashi T. Magnetite nanoparticle-loaded anti-HER2 immunoliposomes for combination of antibody therapy with hyperthermia. *Cancer Lett.* 2004 Aug 30;212(2):167-75.

[37] Chen X, Mao SS. Synthesis of titanium dioxide (TiO2) nanomaterials. *J. Nanosci. Nanotechnol.* 2006 Apr;6(4):906-25.

[38] Peralta-Hernandez JM, Manriquez J, Meas-Vong Y, Rodriguez FJ, Chapman TW, Maldonado MI, Godinez LA. Photocatalytic properties of nano-structured TiO(2)-carbon films obtained by means of electrophoretic deposition. *J. Hazard Mater.* 2007;147:588-93.

[39] Braun V, Hantke K, Koster W. Bacterial iron transport: mechanisms, genetics, and regulation. *Met. Ions Biol. Syst.* 1998;35:67-145.

[40] Kogan MJ, Olmedo I, Hosta L, Guerrero AR, Cruz LJ, Albericio F. Peptides and metallic nanoparticles for biomedical applied usages. *Nanomed.* 2007 Jun;2(3):287-306.

[41] Hawker CJ, Wooley KL. The convergence of synthetic organic and polymer chemistries. *Science.* 2005 Aug 19;309(5738):1200-5.

[42] Bochenkov VE, Sergeev GB. Preparation and chemiresistive properties of nanostructured materials. *Adv. Colloid Interface Sci.* 2005 Nov 30;116(1-3):245-54.

[43] Harada-Shiba M, Yamauchi K, Harada A, Takamisawa I, Shimokado K, Kataoka K. Polyion complex micelles as vectors in gene therapy--pharmacokinetics and in vivo gene transfer. *Gene Ther.* 2002 Mar;9(6):407-14.

[44] Zhu J, Tang A, Law LP, Feng M, Ho KM, Lee DK, Harris FW, Li P. Amphiphilic core-shell nanoparticles with poly(ethylenimine) shells as potential gene delivery carriers. *Bioconjug Chem.* 2005 Jan-Feb;16(1):139-46.

[45] Ai H, Jones SA, Lvov YM. Biomedical applied usages of electrostatic layer-by-layer nano-assembly of polymers, enzymes, and nanoparticles. *Cell Biochem. Biophys.* 2003;39(1):23-43.

[46] Pinto Reis C, Neufeld RJ, Ribeiro AJ, Veiga F. Nanoencapsulation I. Methods for preparation of medicine-loaded polymeric nanoparticles. *Nanomedicine.* 2006 Mar;2(1):8-21.

[47] Pinto Reis C, Neufeld RJ, Ribeiro AJ, Veiga F. Nanoencapsulation II. Biomedical applied usages and present status of peptide and protein nanoparticulate delivery systems. *Nanomedicine.* 2006 Jun;2(2):53-65.

[48] Mayer C. Nanocapsules as medicine delivery systems. *Int. J. Artif. Organs.* 2005 Nov;28(11):1163-71.

[49] Jain KK. Nanotechnology-based medicine delivery for cancer. *Technol. Cancer Res. Treat.* 2005 Aug;4(4):407-16.

[50] Kisak ET, Coldren B, Evans CA, Boyer C, Zasadzinski JA. The vesosome-- a multicompartment medicine delivery vehicle. *Curr. Med. Chem.* 2004 Jan;11(2):199-219.

[51] Baby Y. Nanotechnology in medicine. *Nippon Rinsho.* 2006 Feb;64(2):189-98.

[52] Beck RC, Pohlmann AR, Guterres SS. Nanoparticle-coated microparticles: preparation and characterization. *J. Microencapsul.* 2004 Aug;21(5):499-512.

[53] Monteiro OC, Neves MC, Trindade T. Zinc sulfide nanocoating of silica submicron shperical objects using a single-source method. *J. Nanosci. Nanotechnol.* 2004 Jan-Feb;4(1-2):146-50.

Chapter 3

NANOPOLYMER AND NANOPOLYMERIZATION

WHAT IS A NANOPOLYMER? [1 - 2]

Polymer is a complex generated from a monomer. It means that a polymer is a complex periled that consists of more than one molecule. The polymer can be seen in multiple sizes. It can be said that almost everything we use in daily life has a polymer nature. Various polymerization parameters are mentioned including concentration of monomer and initiator, type and concentration of surfactant, temperature and time of polymerization, ionic strength, pH and co-solvent concentration [3]. These parameters impact on the properties, including size and size distribution, yield and stability of the particles [3]. In nanoscience, polymers also play an important role. The polymer in nanolevel is called nanopolymer, which possesses the properties of nanomaterial: quantum effect, surface effect and interface effect. Basically, many different ways are employed in the preparation of polymer micro- and nanoparticles from monomers. The details on each important way will be further discussed.

1. Emulsion Polymerization

Emulsion polymerization brings particles of about 50-200 nm in diameter. This method produces regular, spherical particles.

Table 1. Some interesting reports concerning emulsion polymerization for nanoparticles

Authors	Details
Joumaa et al. [4]	In this work, submicrometer fluorescent polystyrene particles have been synthesized via miniemulsion polymerization using CdSe/ZnS core-shell quantum dots [4]. In addition, the influence of quantum dot concentration, quantum dot coating (either trioctylphosphine oxide (TOPO)-coated or vinyl-functionalized), and surfactant concentration on the polymerization kinetics and the photoluminescence properties of the treated particles was analyzed [4]. According to this work, Joumaa et al. concluded that the obtained fluorescent particles helped open up new chances for a multiple appearance of applied usages in biotechnology [4].

Table 1. Continued

Authors	Details
Zhang et al. [5]	In this work, the polystyrene latex nanoparticles bearing silanol groups on their surfaces were completely synthesized via miniemulsion polymerization using gamma-methacryloxypropyltrimethoxysilane as the functional comonomer and oil-soluble AIBN as the initiator at neutral conditions [5]. Zhang et al. reported that these silanol-functionalized latex particles could be easily coated with silica or other inorganic or organic compounds to prepare new hybrid particles and hollow microshperical objects [5].
Tan and Tong [6]	Tan and Tong performed an interesting experiment on using different structure types of microemulsions for the preparation of poly(alkylcyanoacrylate) nanoparticles by interfacial polymerization [6]. Tan and Tong reported that the entrapment of the protein within these particles was up to 95%, depending on the summation of monomer used for polymerization and the type of microemulsion used as a polymerization template [6].

2. Emulsifier Free Emulsion Polymerization

Emulsifier free emulsion polymerization produces particles of about 100-1000 nm (0.1-1.0 micron). This method produces regular, spherical particles.

Table 2. Some interesting reports concerning emulsifier free emulsion polymerization for nanoparticles

Authors	Details
Ngai and Wu [7]	Ngai and Wu reported on double roles of stabilization and destabilization of initiator potassium persulfate in surfactant-free emulsion polymerization of styrene under microwave irradiation [7]. In this work, emulsifier-free emulsion polymerization of styrene in a hydro/acetone mixture under microwave irradiation caused narrowly distributed stable polystyrene nanoparticles with an averaged hydrodynamic radius down to 35 nm when 50 wt % of acetone was added [7].
Khvan et al. [8]	In this work, hydrophobic polymer nanoparticles performed through an emulsifier-free emulsion polymerization method were completely integrated into a gallery of pristine sodium montmorillonite via interfacial cation exchange [8]. Khvan et al suggested that, on the basis of the obtained pre-exfoliated clay masterbatch, the existence of strong polymer-clay interactions could improve the mechanical characteristic of nanocomposites [8].
Xu et al. [9]	In this work, a new method to construct a sensitive mediatorless sensor of H2O2 was described [9]. At first, a cleaned gold electrode was immersed in thiol-functionalized poly(styrene-co-acrylic acid) (St-co-AA) nanosphere latex treated by emulsifier-free emulsion polymerization St with AA and function with dithioglycol to assemble the nanoshperical objects, then gold nanoparticles were chemisorbed onto the thiol groups and formed monolayers on the surface of poly(St-co-AA) nanoshperical objects [9]. Finally, horseradish peroxidase (HRP) was immobilized on the surface of the gold nanoparticles [9].

3. Dispersion Polymerization

Dispersion polymerization produces particles in the region of 0.3-10 micron. This method produces regular, spherical particles.

Table 3. Some interesting reports concerning dispersion polymerization for nanoparticles

Authors	Details
Jang et al. [10]	Jang et al reported on fabrication of CdS/PMMA core/shell nanoparticles by dispersion mediated interfacial polymerization [10]. In this work, CdS/PMMA core/shell nanoparticles were synthesized using dispersion-mediated interfacial polymerization [10].
Deng et al. [11]	In this work, Poly(methacrylic acid-grafted-poly(ethylene glycol)) (P(MAA-g-EG)) hydrogel nanoparticles (HNPs) were treated by a thermally-initiated free radical dispersion polymerization method [11]. Deng et al. said that the P(MAA-g-EG) HNPs performed pH-responsive swelling property, which was strongly influenced by the crosslinker dosage [11].
Corbierre et al. [12]	Corbierre et al studied gold nanoparticle/polymer focusing on nanocomposites dispersion of nanoparticles as a function of capping agent molecular weight and grafting density [12]. In this work, the dispersion of polymer-covered gold nanoparticles in high molecular weight polymer matrixes was reported [12]. Corbierre et al. found that dispersion occurred when the matrix molecular weight was greater than that of the brush [12].
Chauvierre et al. [13]	The purpose of this research was to develop a new method which allowed the easy fabrication of nanoparticles with surface properties using polysaccharides [13]. In this work, the preparation was based on the use of a redox radical polymerization reaction applied for the first time to the emulsion polymerization of alkylcyanoacrylates in aqueous continuous media [13].

4. Suspension Polymerization

Suspension polymerization brings about the formation of particles of about 20 micron-2 mm. The gap in the 10-20 micron region may be filled by either seeded polymerization or by more elaborately performed suspension polymerization. This method produces regular, spherical particles.

Table 4. Some interesting reports concerning suspension polymerization for nanoparticles

Authors	Details
Yang et al. [14]	An effective method for purification of nattokinase from fermentation broth using magnetic poly(methyl methacrylate) (PMMA) beads immobilized with p-aminobenzamidine was proposed in this study [14]. Firstly, magnetic PMMA beads with a narrow size distribution were treated by spraying suspension polymerization then, they were highly functionalized via transesterification reaction with polyethylene glycol [14].

Table 4. (Continued)

Authors	Details
Jantarat et al. [15]	In this work, molecularly imprinted polymer (MIP) nanoparticle-on-microspherical objects (NOM) selective for S-propranolol were completely treated using suspension polymerization involving agitation of the reaction mixture at high speed [15]. Jantarat et al said that the proposed MIP-NOM composite membrane controlled release system might be applied used for fabrication of new membranes with self-controllable permeability responding to the existence of target solutes [15].

5. Precipitation Polymerization

Precipitation polymerization gives irregularly shaped particles in the range of 0.1-10 micron. It is useful only if low molecular weights of the polymer and polydispersity of the particles do not adversely impact the intended applied usage of the product.

Table 5. Some interesting reports concerning precipitation polymerization for nanoparticles

Authors	Details
Yoshimatsu et al. [16]	Yoshimatsu reported on uniform molecularly imprinted microspherical objects and nanoparticles treated by precipitation polymerization [16]. Yoshimatsu et al initiated a study on new synthetic conditions to obtain MIP beads with controllable size in the nano- to micro-meter range, using racemic propranolol as a model template [16]. In this work, varying the composition of the cross-linking monomer allowed the particle size of the MIP beads to be altered in the range of 130 nm to 2.4 microm, whereas the designable binding property of the imprinted beads remained intact [16].
Cai et al. [17]	In this work, nearly monodisperse nanoparticles have been synthesized based on a naturally occurring polymer of hydropropylcellulose (HPC) using precipitation polymerization method [17]. Cai et al found that the HPC nanoparticle assembly in hydro could be further stabilized by covalently bonding neighboring particles to form a three-dimensional network [17].

6. Seed Polymerization

Seed polymerization is the polymerization aiming at seeding of particles. This can result in a seeded polymer.

Table 6. Some interesting reports concerning seed polymerization for nanoparticles

Authors	Details
Shamin et al. [18]	In this work, thermosensitive polymer coated nanomagnetic adsorbents were synthesized by seed polymerization using surface modified nanomagnetic particles as the seeds [18]. In detail, the Fe3O4 nanomagnetic particles were treated by chemical precipitation of Fe2+ and Fe3+ salts in the ratio of 1:2 under alkaline and inert condition [18].
Shamin et al. [19]	In this work, adsorption and desorption property of bovine serum albumin on surface-modified magnetic nanoparticles covered with thermosensitive polymer was investigated as a function of temperature, pH, and ionic strength [19]. Functionalization of surface-modified magnetic particles was performed by seed polymerization using N-isopropylacrylamide as the main monomer [19].

7. Anionic Polymerization

Anionic polymerization means polymerization in which the catalysts are Lewis bases, such as alkali metals.

Table 7. Some interesting reports concerning anionic polymerization for nanoparticles

Authors	Details
Arias et al. [20]	In this article a method is described to prepare composite colloidal nanoparticles, consisting of a magnetic core (carbonyl iron) and a biodegradable polymeric shell [poly(butylcyanoacrylate) or PBCA] [20]. The method is based on the so-called anionic polymerization procedure [20].
Gibuad et al. [21]	Gibuad et al studied the polyalkylcyanoacrylate nanoparticles as carriers for granulocyte-colony stimulating factor (G-CSF) [21]. By anionic polymerization, it was probable to associate more than 66% of rhG-CSF with nanoparticles (polyisobutyl- or polyisohexylcyanoacrylate nanoparticles) when the glycoprotein was added at the end of the polymerization process [21]. It could be shown that the rhG-CSF was mainly adsorbed on the surface of the nanoparticles and most of the colony stimulating activity was conserved [21].
Park et al. [22]	New thermosensitive polyion complex (PIC) micelles were treated in an aqueous medium based on the complexation of a pair of oppositely charged block ionomers, poly(2-isopropyl-2-oxazoline)-b-poly(amino acid)s (PiPrOx-b-PAA), consisting thermosensitive PiPrOx segments [22]. The controlled synthesis of PiPrOx-b-PAA was obtained via the ring-opening anionic polymerization of N-carboxyanhydrides (NCA) of either eta-benzyloxycarbonyl-l-lysine (Lys(Z)-NCA) or beta-benzyl-l-aspartate (BLA-NCA) with omega-amino-functionalized PiPrOx macroinitiators and the subsequent deprotection reaction under acidic or basic conditions [22].

8. Cationic Polymerization

Cationic polymerization means polymerization in which the catalysts are Lewis acids, such as acidic metals.

Table 8. Some interesting reports concerning cationic polymerization for nanoparticles

Authors	Details
Kuo et al. [23]	In this work, a new type of surfactant with a hydrophile of dendritic polyethylenimine (C(12)(EI)(7)) was synthesized by a cationic polymerization of dodecylamine with aziridine and was used as a stabilizer to prepare Pt colloid, which is then used in situ to prepare carbon-supported Pt nanoparticles [23]. In addition, the interaction between Pt and C(12)(EI)(7) were studied and were explained in terms of a mechanism of couple-functional stabilization both on carbon and in the thermal treatment [23].

9. Free Radical Polymerization

Free radical polymerization means polymerization in which the catalysts are free radicals.

Table 9. Some interesting reports concerning free radical polymerization for nanoparticles

Authors	Details
Hoffmann et al. [24]	Hoffmann et al reported on preparation, characterization and cytotoxicity of methylmethacrylate copolymer nanoparticles with a permanent positive surface charge [24]. In this work, methylmethacrylate copolymer nanoparticles consisting different cationic comonomers such as N-trimethylammoniumethylmethacrylate (TMAEMC), N-dimethylammoniumethylmethacrylate (DMAEMC), N-trimethylammoniumpropylmethylacrylamide (MAPTAC) or the anionic comonomer sulfopropylmethacrylate (SPM), respectively, were treated by free radical polymerization [24].
Kim et al. [25]	In this work, stable rodlike nanoparticles with highly controlled surface charge density have been developed by the free radical polymerization of the mixture of polymerizable cationic surfactant, cetyltrimethylammonium 4-vinylbenzoate (CTVB), and hydrotropic salt sodium 4-styrenesulfonate (NaSS) in aqueous solution [25].
Feng and Li [26]	Feng and Li performed a study to investigate the properties of cationic nanoparticles composed of poly (ethyleneimine)-g-poly (methyl methacrylate) (PEI-PMMA) nanoparticles were synthesized by free radical polymerization [26]. Feng and Li concluded that PEI-PMMA nanoparticles effectively transferred DNA to target cells and it was a useful non-viral carrier for gene delivery [26].

Dimension of Nanopolymer

The nanopolymer can be a one-dimensional nanopolymer, a two-dimensional nanopolymer or a three-dimensional nanopolymer. The important nanopolymers will be hereby discussed.

1. One-Dimensional Nanopolymer

1.1. Nanoflim

Nanoflim is a very thin film with a thickness of less than 100 nm. Nanofilm is an important part in the development of nanofilm-based disarmer. At present, the field of polypeptide multilayer nanofilm research flourishes where the study of protein structure and function shares a border with development of polyelectrolyte multilayers [27]. The nanofilm is accepted for its usefulness for the fabrication process in the development of a biodevice (Table 10).

Table 10. Some interesting reports concerning nanofilm for development of biochip device

Authors	Details
Yasuda et al. [28]	Yasuda et al. reported on biocompatibility of nanofilm-encapsulated silicone and silicone-hydrogel contact lenses [28]. Yasuda et al. said that the encapsulation of a highly oxygen-permeable contact lens by a nanofilm with an imperturbable surface state decreased the unwanted effects of the contact lens [28].
Srivastava and McShane [29]	In this work, the potential to limit diffusion of macromolecules embedded in alginate spherical objects with nanofilm coatings was assessed [29]. The effectiveness of nanofilms reducing loss of the encapsulated macromolecules was found to vary between different polycation materials used [29]. This study supports the feasibility of using these microsystems for development of long-term stable encapsulated systems, such as implantable biosensors [29].
Srivastava et al. [30]	In this work, multilayer nanofilms of diazo-resin with poly(styrene sulfonate) were used in an effort to stabilize the encapsulation of glucose oxidase enzyme for biosensor applied usages [30]. Srivastava et al. demonstrated that enzyme immobilization and stabilization could be obtained by using simple modifications to the layer-by-layer self-assembly method [30].
Srivastava et al. [31]	Srivastava et al studied the stable encapsulation of an active enzyme by applied usage of multilayer nanofilm coatings to alginate microspherical objects [31]. In this work, layer-by-layer self-assembly was explored as a potential way to bring a diffusion barrier to encapsulated glucose oxidase inside alginate microspherical objects, fabricated using an emulsification method [31].
Mavrokefalos et al. [32]	Four-probe measurements of the in-plane thermoelectric properties of nanofilms were discussed in this work [32]. Mavrokefalos et al. reported a method based on a suspended microdevice for four-probe measurements of the Seebeck coefficient, thermal conductivity, electrical conductivity, and thermoelectric figure of merit of patterned indium arsenide nanofilms integrated on the microdevice [32].

2. Two-Dimensional Nanopolymer

2.1. Carbon Nanotube

Carbon nanotube is useful for development of a biosensor. The carbon nanotube can be divided into two types, single-walled and multiple-walled carbon nanotubes. The basic polymerised dentine of the well of carbon nanotube can be either armchair pattern or zigzag pattern. Chemical attachments to the walls or ends of carbon nanotubes are good [33]. These chemical bonds might be used to tailor the interaction of the nanotube with other entities, such as a solvent, a polymer matrix, or other nanotubes [33]. The carbon nanotube has a very good elasticity and can be bent. Due to the high modules, it is highly stable mechanically. It can act as a heat transferring and electrical transfusing media. It is heat stable. Due to its organic nature, it can be used medically in vivo. Due to its two-dimensional appearance, other tans or molecules can be inserted into the hollow inside the carbon nanotube and this can be the transportation tract.

Carbon nanotube also presents the quantum effect similar to other general nanostructure materials. Field emitter property is the basic quality that brings the carbon nanotube to be an important part in many advanced biosensors. At present, carbon nanotubes are considered one of the most useful materials in nanotechnology, with attractive properties for many technological applied usages [34]. The different synthesis, purification, and postprocessing methods produce CNTs with different physical characteristics, which can be applied in different fields ranging from composite materials, medical applied usages, and electronics to energy storage [34]. The chemically functionalized nanotube might have mechanical or electrical properties that are different from those of the unfunctionalized nanotube and thus might be used as a chemical sensor or a nanometer-scale electronic device [33]. It is also generally used in medical devices [35 - 36].

Table 11. Some interesting reports concerning the carbon nanotube in biodevices

Authors	Details
Kang and Park [37]	In this work, a highly amplified magnetic force in adjoining microchannels was exploited by employing microfluidics and a magnetic-field-induced saw-tooth nickel microstructure [37].
Whitten et al. [38]	Whitten presented new free-standing carbon nanotube composite bio-electrodes [38]. Preliminary studies revealed that the new carbon nanotube biopolymer electrodes could foster prolific L929 cell growth [38].
Huang et al. [39]	In this work, carboxylated single-walled carbon nanotubes (SWCNT) chemically integrated on gold substrate were employed as a netlike electrode to investigate the charge-transfer process and electrode process kinetics using uric acid as an example [39]. The results indicated that an electrode processed in the carboxylated SWCNT electrode system was governed by the surface adsorption-controlled electrochemical process [39].
Gudiksen et al. [40]	Gudiksen et al. reported on growth of nanowire superlattice structures for nanoscale photonics and electronics [40]. According to this work, single-nanowire photoluminescence, electrical transport and electroluminescence measurements revealed the unique photonic and electronic properties of these nanowire superlattices, and suggested potential applied usages ranging from nano-barcodes to polarized nanoscale light-emitting diodes [40].
Meng et al. [41]	In this work, the adsorption property of albumin, fibrinogen, and fresh plasma on single-walled carbon nanotube membranes was studied using scanning electron microscopy, energy-dispersive x-ray, and direct enzyme-linked immunosorbent assay, respectively [41]. Meng et al. found that the function of adsorbed fibrinogen to mediate platelet recognition, adhesion, activation, and aggregation was significantly suppressed, which induced extremely low levels of platelet adhesion and activation [41].

Authors	Details
Bottini et al. [42]	Bottini et al. reported the formation of a supramolecular luminescent nanoassembly composed of indivicouple or small ropes of full-length, single-walled carbon nanotubes decorated with streptavidin-conjugated quantum dots [42]. Bottini et al. said that the nanoassembly was found to be transported to lysosomes once internalized by cells [42].
Meng et al. [43]	In this study, a new type of carbon nanotube material, so-called nonwoven single-walled carbon nanotubes with nanotopographic structure and macroscopic volume, was used as a cell-growing scaffold [43]. The results obtained in this work are of significance to in vitro cell amplification in large-scale tissue regeneration, guided repair, and as a biomedical device applied usage [43].
Zhou et al. [44]	Zhou et al. reported on bacteria capture, concentration and detection by alternating present dielectrophoresis and self-assembly of dispersed single-wall carbon nanotubes [44]. In this work, the high polarizability and dielectrophoretic mobility of single-walled carbon nanotubes were utilized to capture and detect low numbers of bacteria and submicron particles in milliliter-sized samples [44].

3. Three-Dimensional Nanopolymer

3.1. Liposome [45 - 47]

Liposome is a spherical-shaped nanopolymer. The average diameter of liposome is about 100 nm. The main composition of liposome is sugar and protein that can act as the receptor. Liposome is generally used as a medicine delusory system and the goal is to use liposome as a magic bullet. Functionalized liposome is liposome that consists of a functional molecule inside, while it is mostly a medicine. Functionalized liposome is presently studied for cancer therapy. At present, multifunctional liposome has been completely developed. Based on the antigen-antibody reaction, multifunctional liposome can be a better medicine carrier to the designed target organ. A good example is the multifunctical liposome consisting doxyrubicin for treatment of ovarian cancer. Due to the nature of liposome being a lipid, it can easily penetrate into skin but has low stability. During the last three decades, the value of liposomes as a medicine delivery system has been examined [47]. The interest in liposomes as carriers of active substances is based on their potential to enclose various types of biological materials and to deliver them to diverse cell types [47]. More details on applied usages of liposome in a pharmacological aspect can be seen in another chapter in this book.

3.2. Dendrimer [48 - 51]

Dendrimer is another important spherical nanopolymer. Dendrimer is a synthesized nanopolymer with dendrogram, three-branched style, structure. The dendrogram structure of dendrimer is equally distributed in every plane and the result is a symmetrical spherical structure with very high surface area and high stability. More details on applied usages of dendrimer in a pharmacological aspect can be seen in another chapter in this book.

Figure 1. Diagram showing cut surface of dendrimer.

Due to the interior branching dendrogram pattern, dendrimer is more stable than liposome. Similar to liposome, dendrimer can serve as a nanoscale platform for creating a multifunctional device capable of detection and therapy. Dendrimer with these properties requires surface modification and so-called functionalized tectodendrimer. Multifunctional tectoderndrimer is also available. The G8 dendrimer is the best example of multifunctional tectodendrimer [52].

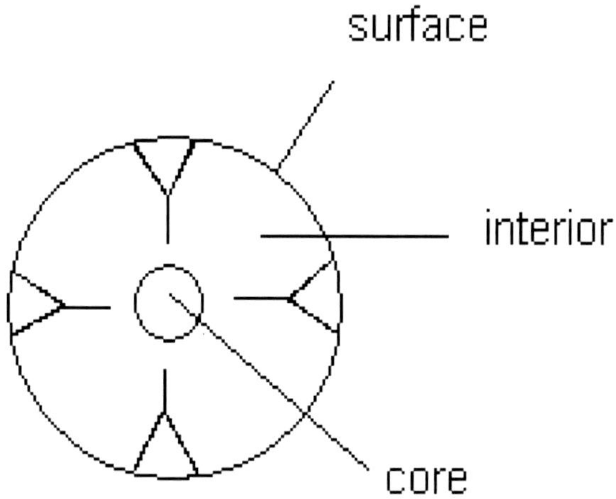

3.3. Nanoshell [53 - 54]

Nanoshell is another generalized use nanopolymer. Nanoshell is a spherical nanopolymer with gold shell and interior dielectric substance such as silicon dioxide. This combination causes the change in quantum effect. Nanoshell has a new property, that of absorbing near infra-red electromagnetic wave instead of visible light. This makes it useful as a new cytodestructive medicine via hyperthermia mechanism.

3.4. Polyethyleneimine (PEI) [55]

PEI is a nanoparticle that is commonly used in gene transferring. PEI is proved to be a good gene transfer vehicle. However, an important consideration on PEI is its probable carcinogenicity [56]. Therefore, PEI is given Chertier limited use at present.

3.5. Gold-PEI [55]

Gold-PEI is the combination of PEI and gold nanoparticle. Presently the use of this combination is also limited due to the concern regarding PEI as previously described.

3.6. Fullerene [57 - 58]

Fullerene of baggy ball is a new nanopolymer that is very useful for medicine carrying. Fullerene has a spherical shape with a surface that is allowable for modification. The modification can be tagged with specific chemicals that can result in therapeutic property. The modification can be performed because fullerene has an interesting surface structure in that it possesses many one-nm-size spaces that allow binding of another chemical. At present, many new medicines are developed based on the fullerene. A good example is fullerene-based inhibitor of the HIV-1 protease. More details on applied usages of fullerene in pharmacological aspect can be seen in another chapter in this book.

REFERENCES

[1] Arshady R. Preparation of polymer nano- and microshperical objects by vinyl polymerization ways. *J. Microencapsul.* 1988 Apr-Jun;5(2):101-14.
[2] Arshady R. Preparation of nano- and microshperical objects by polycondensation ways. *J. Microencapsul.* 1989 Jan-Mar;6(1):1-12.
[3] Boguslavsky L, Baruch S, Margel S. Synthesis and characterization of polyacrylonitrile nanoparticles by dispersion/emulsion polymerization process. *J. Colloid. Interface Sci.* 2005 Sep 1;289(1):71-85.
[4] Joumaa N, Lansalot M, Theretz A, Elaissari A, Sukhanova A, Artemyev M, Nabiev I, Cohen JH. Synthesis of quantum dot-tagged submicrometer polystyrene particles by miniemulsion polymerization. *Langmuir.* 2006 Feb 14;22(4):1810-6.
[5] Zhang SW, Zhou SX, Weng YM, Wu LM. Synthesis of silanol-functionalized latex nanoparticles through miniemulsion copolymerization of styrene and gamma-methacryloxypropyltrimethoxysilane. *Langmuir.* 2006 May 9;22(10):4674-9.
[6] Krauel K, Davies NM, Hook S, Rades T. Using different structure types of microemulsions for the preparation of poly(alkylcyanoacrylate) nanoparticles by interfacial polymerization. *J. Control Release.* 2005 Aug 18;106(1-2):76-87.
[7] Ngai T, Wu C. Double roles of stabilization and destabilization of initiator potassium persulfate in surfactant-free emulsion polymerization of styrene under microwave irradiation. *Langmuir.* 2005 Aug 30;21(18):8520-5.
[8] Khvan S, Kim J, Lee SS. Mechanistic examination of pre-exfoliating confinement of surface-active polystyrene nanobeads within pristine clay. *J. Colloid Interface Sci.* 2007 Feb 1;306(1):22-7.
[9] Xu S, Tu G, Peng B, Han X. Self-assembling gold nanoparticles on thiol-functionalized poly(styrene-co-acrylic acid) nanoshperical objects for fabrication of a mediatorless biosensor. *Anal. Chim. Acta.* 2006 Jun 16;570(2):151-7.
[10] Jang J, Kim S, Lee KJ. Fabrication of CdS/PMMA core/shell nanoparticles by dispersion mediated interfacial polymerization. *Chem. Commun* (Camb). 2007 Jul 14;(26):2689-91
[11] Deng L, He X, Li A, Yang Q, Dong A. Preparation of poly(MAA-g-EG) hydrogel nanoparticles by a thermally-initiated free radical dispersion polymerization. *J. Nanosci. Nanotechnol.* 2007 Feb;7(2):626-33.
[12] Corbierre MK, Cameron NS, Sutton M, Laaziri K, Lennox RB. Gold nanoparticle/polymer nanocomposites: dispersion of nanoparticles as a function of capping agent molecular weight and grafting density. *Langmuir.* 2005 Jun 21;21(13):6063-72.
[13] Chauvierre C, Vauthier C, Labarre D, Couvreur P, Marden MC, Leclerc L. A new generation of polymer nanoparticles for medicine delivery. *Cell Mol. Biol.* (Noisy-le-grand). 2004 May;50(3):233-9.
[14] Yang C, Xing J, Guan Y, Liu H. Superparamagnetic poly(methyl methacrylate) beads for nattokinase purification from fermentation broth. *Appl. Microbiol. Biotechnol.* 2006 Sep;72(3):616-22.
[15] Jantarat C, Tangthong N, Songkro S, Martin GP, Suedee R. S-Propranolol imprinted polymer nanoparticle-on-microsphere composite porous cellulose membrane for the

enantioselectively controlled delivery of racemic propranolol. *Int. J. Pharm.* 2007 Jul 31;

[16] Yoshimatsu K, Reimhult K, Krozer A, Mosbach K, Sode K, Ye L. Uniform molecularly imprinted microshperical objects and nanoparticles treated by precipitation polymerization: the control of particle size suitable for different analytical applied usages. *Anal. Chim. Acta.* 2007 Feb 12;584(1):112-21.

[17] Cai T, Hu Z, Marquez M. Synthesis and self-assembly of nearly monodisperse nanoparticles of a naturally occurring polymer. *Langmuir.* 2004 Aug 31;20(18):7355-9.

[18] Shamim N, Hong L, Hidajat K, Uddin MS. Thermosensitive polymer (N-isopropylacrylamide) coated nanomagnetic particles: preparation and characterization. *Colloids Surf. B Biointerfaces.* 2007 Mar 15;55(1):51-8.

[19] Shamim N, Hong L, Hidajat K, Uddin MS. Thermosensitive-polymer-coated magnetic nanoparticles: adsorption and desorption of bovine serum albumin. *J. Colloid Interface Sci.* 2006 Dec 1;304(1):1-8.

[20] Arias JL, Gallardo V, Linares-Molinero F, Delgado AV. Preparation and characterization of carbonyl iron/poly(butylcyanoacrylate) core/shell nanoparticles. *J. Colloid Interface Sci.* 2006 Jul 15;299(2):599-607.

[21] Gibaud S, Rousseau C, Weingarten C, Favier R, Douay L, Andreux JP, Couvreur P. Polyalkylcyanoacrylate nanoparticles as carriers for granulocyte-colony stimulating factor (G-CSF). *J. Control. Release.* 1998 Mar 2;52(1-2):131-9.

[22] Park JS, Akiyama Y, Yamasaki Y, Kataoka K. Preparation and characterization of polyion complex micelles with a new thermosensitive poly(2-isopropyl-2-oxazoline) shell via the complexation of oppositely charged block ionomers. *Langmuir.* 2007 Jan 2;23(1):138-46.

[23] Kuo PL, Chen WF, Huang HY, Chang IC, Dai SA. Stabilizing effect of pseudo-dendritic polyethylenimine on platinum nanoparticles supported on carbon. *J. Phys. Chem. B.* 2006 Feb 23;110(7):3071-7.

[24] Hoffmann F, Cinatl J, Kabickova H, Kreuter J, Stieneker F. Preparation, characterization and cytotoxicity of methylmethacrylate copolymer nanoparticles with a permanent positive surface charge. *Int. J. Pharm.* 1997 Nov 28;157(2):189- 198.

[25] Kim TH, Choi SM, Kline SR. Polymerized rodlike nanoparticles with controlled surface charge density. *Langmuir.* 2006 Mar 14;22(6):2844-50.

[26] Feng M, Li P. PEI-PMMA cationic nanoparticles as carriers for gene transfer. Yao Xue Xue Bao. 2005 Oct;40(10):893-7.

[27] Haynie DT, Zhang L, Zhao W, Rudra JS. Protein-inspired multilayer nanofilms: science, technology and medicine. *Nanomedicine.* 2006 Sep;2(3):150-7.

[28] Yasuda H. Biocompatibility of nanofilm-encapsulated silicone and silicone-hydrogel contact lenses. *Macromol. Biosci.* 2006 Feb 10;6(2):121-38.

[29] Srivastava R, McShane MJ. Applied usage of self-integrated ultra-thin film coatings to stabilize macromolecule encapsulation in alginate microshperical objects. *J. Microencapsul.* 2005 Jun;22(4):397-411.

[30] Srivastava R, Brown JQ, Zhu H, McShane MJ. Stabilization of glucose oxidase in alginate microshperical objects with photoreactive diazoresin nanofilm coatings. *Biotechnol. Bioeng.* 2005 Jul 5;91(1):124-31.

[31] Srivastava R, Brown JQ, Zhu H, McShane MJ. Stable encapsulation of active enzyme by applied usage of multilayer nanofilm coatings to alginate microshperical objects. *Macromol. Biosci.* 2005 Aug 12;5(8):717-27.

[32] Mavrokefalos A, Pettes MT, Zhou F, Shi L. Four-probe measurements of the in-plane thermoelectric properties of nanofilms. *Rev. Sci. Instrum.* 2007 Mar;78(3):034901.

[33] Sinnott SB. Chemical functionalization of carbon nanotubes. *J. Nanosci. Nanotechnol.* 2002 Apr;2(2):113-23.

[34] Boczkowski J, Lanone S. Potential uses of carbon nanotubes in the medical field: how worried should patients be? *Nanomed.* 2007 Aug;2(4):407-10.

[35] Helland A, Wick P, Koehler A, Schmid K, Som C. Reviewing the environmental and human health knowledge base of carbon nanotubes. *Environ. Health Perspect.* 2007 Aug;115(8):1125-31.

[36] Koyama S, Haniu H, Osaka K, Koyama H, Kuroiwa N, Endo M, Kim YA, Hayashi T. Medical applied usage of carbon-nanotube-filled nanocomposites: the microcatheter. *Small.* 2006 Dec;2(12):1406-11.

[37] Kang JH, Park JK. Magnetophoretic continuous purification of single-walled carbon nanotubes from catalytic impurities in a microfluidic device. *Small.* 2007 Oct;3(10):1784-91.

[38] Whitten PG, Gestos AA, Spinks GM, Gilmore KJ, Wallace GG. Free standing carbon nanotube composite bio-electrodes. *J. Biomed. Mater Res. B Appl. Biomater.* 2007 Jul;82(1):37-43.

[39] Huang XJ, Im HS, Yarimaga O, Kim JH, Lee DH, Kim HS, Choi YK. Direct electrochemistry of uric acid at chemically integrated carboxylated single-walled carbon nanotubes netlike electrode. *J. Phys. Chem. B.* 2006 Nov 2;110(43):21850-6.

[40] Gudiksen MS, Lauhon LJ, Wang J, Smith DC, Lieber CM. Growth of nanowire superlattice structures for nanoscale photonics and electronics. *Nature.* 2002 Feb 7;415(6872):617-20.

[41] Meng J, Song L, Xu H, Kong H, Wang C, Guo X, Xie S. Effects of single-walled carbon nanotubes on the functions of plasma proteins and potentials in vascular prostheses. *Nanomedicine.* 2005 Jun;1(2):136-42.

[42] Bottini M, Cerignoli F, Dawson MI, Magrini A, Rosato N, Mustelin T. Full-length single-walled carbon nanotubes decorated with streptavidin-conjugated quantum dots as multivalent intracellular fluorescent nanoprobes. *Biomacromolecules.* 2006 Aug;7(8):2259-63.

[43] Meng J, Song L, Meng J, Kong H, Zhu G, Wang C, Xu L, Xie S, Xu H. Using single-walled carbon nanotubes nonwoven films as scaffolds to enhance long-term cell proliferation in vitro. *J. Biomed. Mater Res. A.* 2006 Nov;79(2):298-306.

[44] Zhou R, Wang P, Chang HC. Bacteria capture, concentration and detection by alternating present dielectrophoresis and self-assembly of dispersed single-wall carbon nanotubes. *Electrophoresis.* 2006 Apr;27(7):1376-85.

[45] Fang JY. Nano- or submicron-sized liposomes as carriers for medicine delivery. *Chang. Gung. Med J.* 2006 Jul-Aug;29(4):358-62.

[46] Ranade VV. Medicine delivery systems. 1. site-specific medicine delivery using liposomes as carriers. *J. Clin. Pharmacol.* 1989 Aug;29(8):685-94.

[47] Schmid MH, Korting HC. Liposomes: a medicine carrier system for topical treatment in dermatology. *Crit Rev. Ther. Medicine Carrier Sys*t. 1994;11(2-3):97-118.

[48] Singh P. Dendrimers and their applied usages in immunoassays and clinical diagnostics. *Biotechnol. Appl. Biochem.* 2007 Sep;48(Pt 1):1-9.

[49] Gupta U, Agashe HB, Asthana A, Jain NK. A review of in vitro-in vivo investigations on dendrimers: the new nanoscopic medicine carriers. *Nanomedicine.* 2006 Jun;2(2):66-73

[50] Bai S, Thomas C, Rawat A, Ahsan F. Recent progress in dendrimer-based nanocarriers. *Crit. Rev. Ther. Medicine Carrier Syst.* 2006;23(6):437-95.

[51] Yang H, Kao WJ. Dendrimers for pharmaceutical and biomedical applied usages. *J. Biomater. Sci. Polym. Ed.* 2006;17(1-2):3-19.

[52] Maiti PK, Goddard WA 3rd. Solvent quality changes the structure of G8 PAMAM dendrimer, a disagreement with some experimental interpretations. *J. Phys. Chem. B.* 2006 Dec 28;110(51):25628-32.

[53] Hirsch LR, Gobin AM, Lowery AR, Tam F, Drezek RA, Halas NJ, West JL. Metal nanoshells.*Ann. Biomed. Eng.* 2006 Jan;34(1):15-22.

[54] Loo C, Lin A, Hirsch L, Lee MH, Barton J, Halas N, West J, Drezek R. Nanoshell-enabled photonics-based imaging and therapy of cancer. *Technol. Cancer Res. Treat.* 2004 Feb;3(1):33-40.

[55] Allen TM, Cheng WW, Hare JI, Laginha KM. Pharmacokinetics and pharmacodynamics of lipidic nano-particles in cancer. *Anticancer Agents Med. Chem.* 2006 Nov;6(6):513-23.

[56] Rigdon RH. Plastics and carcinogenesis. *South Med. J.* 1974 Dec;67(12):1459-65.

[57] Satoh M, Takayanagi I. Pharmacological studies on fullerene (C60), a new carbon allotrope, and its derivatives. *J. Pharmacol. Sci.* 2006;100(5):513-8.

[58] Bosi S, Da Ros T, Spalluto G, Prato M. Fullerene derivatives: an attractive tool for biological applied usages. *Eur. J. Med. Chem.* 2003 Nov-Dec;38(11-12):913-23.

Chapter 4

MICROFLUIDICS, NANOFLUIDICS AND NANOBIOSENSOR

INTRODUCTION TO BIOCHIP TECHNOLOGY

With the start of the development in the 1980s of tiny chip chemical sensors that can easily be mounted on the tip of a catheter, it was thought that clinical chemistry would enter a new phase: on-site service. Continuous in vivo monitoring of many important blood variables is expected to replace the classical method with off-line laboratory determinations. The fusion of microelectronics and molecular biology has created a new technology (microelectronic miniaturization), which brings a rapid, efficient, and cost-effective tool in molecular diagnostics at a high-sample throughput [1]. The biochip has recently been selected as one of the ten scientific highlights in the year 1998 [1]. The applied usage of microelectronics ranges from the polymerase chain reaction (PCR), nucleotide sequence analysis via DNA-chips or capillary electrophoresis-chips to gene expression analysis [1]. In the biochip, flow splitting microchannels, chaotic micromixers, reaction microchambers and detection microfilters are completely integrated. Many biochips are assuming an important role in molecular diagnostics, and their applied usage in point-of-care diagnosis is expected to help the development of personalized medicine [2]. In this article, a summary of the developmental process and important new types of biochips is made and presented.

DEVELOPMENTAL PROCESS OF A BIOCHIP

A. Fabrication

Fabrication processes become a significant factor and prompt the first consideration in biochip design. The fabrication of this memory system relies on the self-assembly of the nucleic acid junction system, which acts as the scaffolding for a molecular wire consisting of polyacetylene-like units [3]. Microstructures on polymer substrate have been constructed and integrated by using many simple technologies including laser-guided direct writing, injection moulding-thermal bonding and hot-embossing-UV adhesive bonding. Laser-induced optical forces can be used to guide and deposit 100 nm - 10 microm-diameter particles onto solid surfaces in a process called 'laser-guided direct writing' [4]. Nearly any particulate material, including both biological and electronic materials, can be manipulated and deposited on surfaces with micrometer accuracy [4]. The microinjection moulding of copolymer and thermal bonding of injection moulded polymer substrates can be applied for mass production [5]. Passive elements can be hot embossed into the polymer microchips, and then the chips can be integrated into a three-dimensional architecture with the interconnect fabricated from an elastomer to produce a leak-free connection between the biochips [6]. The array can be then produced using a photomodification process, which involves three steps: (a) UV exposure of the polymer surface, (b) coupling and (c) washing of the surface [6]. A molecular switch to control is described, which is based on the formation of a charge-transfer complex [3]. A molecular-scale bit is presented which is based on oxidation-reduction potentials of metal atoms or clusters [3]. The readable 'bit' which can be made of these components has a volume of 3 x 10(7) A3, and should operate at electronic speeds over short distances [3].

B. Microfluidic Control

The control of microfluidics in a device has been one of the most challenging issues. Among all kinds of body fluidics, the blood, which delivers the beneficial ingredients to individual organs and carries away the by-products, consists of abundant health-related information and hence becomes the most interesting target in clinical diagnostic applied usages. The flow principle is described as a valveless, electroosmotically driven technology used for controlling the stream profile in a laminar flow chamber [7]. Basically, a blood sample has to pass three steps within a biochip: (a) sample loading, (b) flowing for analysis and (c) explusion of already analyzed sample. Modification of the flow of electroosmotically controlled guiding stream, running parallel to a central sample stream, can be used for positioning the sample stream in the dimension perpendicular to the flow direction [7]. The analyzed sample is loaded from the sample inlet port to the detection chamber by capillary force, without any external intervening forces [8]. For this and to control the time duration of sample fluid in each compartment of the device, including the inlet port, diffusion barrier, reaction chamber, flow-delay neck, and detection chamber, the fluid conduit has been designed with various geometries of channel width, depth, and shape [8]. Especially, the fluid path has been designed so that the sample flow naturally stops after filling the detection chamber to allow sufficient time for biochemical reaction and subsequent washing steps [8]. The feasibility of creating nanometer scale depressions in biological substrates using active enzymes delivered with scanning probe microscopes has been previously demonstrated and provides a simple way to precisely control fluid flow dimension [9]. Such nanochannels may be effectively used in nanofluidic biochip applied usages [9].

C. Amperometric Measurement

The basic principle of the on-chip biosensor array is based on amperometric measurements. Transmembrane flux monitoring is used and further interpreted as laboratory results. Basically, flux change due to the biochemical reaction is temporary and can be amperometrically recorded by biochip microelectrodes [10]. Electrical biochips made in Si-technology portable devices have been constructed for field applied usages and point-of-care diagnosis [11]. These miniaturized amperometric biosensor devices enable the evaluation of biomolecular interactions by measuring the redox recycling of ELISA products, and as the electrical monitoring of metabolites [11]. The highly sensitive redox recycling is helped by interdigitated ultramicroelectrodes of high spatial resolution [11].

IMPORTANT NEW TYPES OF BIOCHIPS

A. GeneChip

GeneChip is a system developed by Affymetrix USA. It employs a specific method for spotting DNA probes on chips, which is different from any other DNA chips, and can complete the whole process from sample preparation to data construction and analysis [12]. The GeneChip system can be applied to both gene expression analysis and genomic mutation analysis, which would play an important role in human genome analysis in the future [12]. DNA microarray technology, especially the use of GeneChip microarrays, has become a standard tool for parallel gene expression analysis [13]. Recent improvements in GeneChip microarrays enable whole-genome expression analysis [13]. GeneChips can be used for high-throughput mutation detection, SNP genotyping, expression profiling and detection of chromosomal aberrations [14]. This opens the way for clinical applied usages in genetics, cytogenetics, pharmacogenetics, oncology and germ recognition [14]. Establishing standards is a central issue in improving data quality and, in combination with automatic, easy-to-interpret reports, will form the basis of the clinical applicability [14].

B. Cyp450

Pharmacogenetics has assumed increasing importance with the producing concepts of personalized medicine. There is a need to determine the metabolic status of an individual when using medicines, the actions of which are influenced by medicine-metabolizing enzymes. CYP450 is an enzyme quoted for the correlation to pharmacodynamics of many medicines. Laboratory determination for CYP450 has now reached its advanced phase, biochip technology. The most well-known chip for CYP450 is AmpliChip CYP450 [15]. The AmpliChip CYP450 Test is based on microarray technology, which combines hybridization in precise sites on a glass microarray and a fluorescent tagging system [16]. AmpliChip CYP450 (Roche Molecular Diagnostics, Alameda, CA, USA) is the first approved microarray molecular diagnostic test for the analysis of 29 polymorphisms and mutations of the CYP2D6 gene, and two polymorphisms of the CYP2C19 gene [15]. It combines Roche's PCR technology with the GeneChip microarray system (Affymetrix, Santa Clara, CA, USA) [15]. The role of CYP450 chip for testing for CYP polymorphisms in the therapeutic use of many medicines can be expected [15 - 17].

C. Protein Biochips

Classical methods of protein analysis such as electrophoresis, ELISA and liquid chromatography are generally time-consuming, labor-intensive and lack high-throughput capacity [18]. In addition, all existing methods used to measure proteins necessitate multiple division of the starting point sample and individual tests carried out for each substance, with an associated cost for each test [18]. Proteomics' requirement for simultaneous measurement of multiple markers is now probable with biochip array technology [19]. Many laboratories utilise in-house, manual procedures for biochip fabrication and sample testing [19]. Reproducibility and standardisation of biochip processes is vital to ensure quality of results and offer the best tool for elucidation of complex relationships between multiple proteins in diseased conditions [19]. The chip system allows many tests to be performed simultaneously without dividing the starting point patient sample [18]. This system helps the development of multiplexed assays that simultaneously measure many different analytes in a small sample volume [18]. These emerging technologies fall into two categories: 1) spotted array-based tools, and 2) microfluidic-based tools [18].

D.Nanotechnology-Based Biochips

There are many attempts to produce new biochips based on new nanotechnology. The resonance-amplified absorption (REA) by metal clusters on a surface is an effective way on which to base bio-optical devices [20]. A four-layer device consisting of a metal mirror, a polymer or glass-type distance layer, a biomolecule interaction layer and a sub-monolayer of biorecognitively bound metal nano-clusters is made and applied for biochips [20]. In addition, polymer microlenses have been manufactured by delivering droplets of a monomer mixture to a glass substrate using a nano fountain pen (NFP) [21]. Subsequent UV polymerization yielded microlenses with optical properties that were controlled by varying the deposition time of the monomer solution [21]. Using this approach, it is probable to strategically place single microlenses at predefined positions with very high accuracy, an ability which may prove very useful for nano-biochip applied usages [21]. Accompanied with the nanolithography method, nanochannel biochip can be completely fabricated by NFP [9, 22].

Principle and Applied Usage of Microfluidics

As previously mentioned, microfluidics is an important basic technology for producing diagnostic devices. The control of microvolume of fluid via the constructed micropath is the basic principle of microfluidics. At present, microfluidic platforms can enable the miniaturization, integration, and automation of biochemical assays [23]. Nowadays, there is nearly an unmanageable multiple of appearances of alternative approaches that can do this in principle [23]. Microfluidic unit operations can be easily combined within a well-defined and cost efficient fabrication technology to implement applied usage specific biochemical assays in an easy, flexible and ideally monolithical way [23]. There are many types of microfluidic devices. Important types are hereby presented.

A. Microfluidic Enzyme-Linked Immunosorbent Assay Technology

A microfluidic enzyme-linked immunosorbent assay is completely developed by first introducing the principle of immunoassay, enzyme-linked immunosorbent assay, and microfabricated devices, followed by a combination of microfabrication technology and the characterization of microfluidic components [24]. The details from important reports on microfluidic enzyme-linked immunosorbent assay are presented in Table 1.

Table 1. Some reports on microfluidic enzyme-linked immunosorbent assay technology

Authors	Details
Lai et al. [25]	Lai et al. presented an integrated microfluidic device on a compact disk that runs an enzyme-linked immunosorbent assay for rat IgG from a hybridoma cell culture [25]. Lai et al. revealed that the microchip-based enzyme-linked immunosorbent assay had the same detection range as the conventional method on the 96-well microtiter plate but had usefulness such as less reagent consumption and shorter assay time over the conventional method [25].
Morier et al. [26]	A way of using gravity flow to induce a linear convection within a microfluidic system was presented in this study [26]. Morier et al. revealed and mathematically supported that tilting a 1 cm long covered microchannel was enough to generate flow rates up to 1000 nL/min, which represents a linear velocity of 2.4 mm/sec [26].
Bai et al. [27]	In this work, a new surface treatment method using polyethyleneimine (PEI), an amine-bearing polymer, was developed to enhance antibody binding on the poly(methyl methacrylate) (PMMA) microfluidic immunoassay device. By treating the PMMA surface of the microchannel on the microfluidic device with PEI, Bai et al. found that 10 times more active antibodies could be bound to the microchannel surface as compared to those without treatment or treated with the small amine-bearing molecule, hexamethylenediamine (HMD) [27].
Moorthy et al. [28]	A colorimetric, disposable botulinum toxin enzyme-linked immunosorbent assay system was reported in this work [28]. For detection of the toxoid, biotinylated antibodies were immobilized on streptavidin-functionalized agarose gel beads [28].
Hermann et al. [29]	In this work, Hermann et al. described a new microfluidic design for the realization of parallel enzyme-linked immunosorbent assay in stop-flow conditions [29]. Magnetic beads were used both as a solid phase to support the formation of the reactive immune complex and to get a magnetic mixing inside the channels [29].

Table 1. Continued

Authors	Details
Hoegger et al. [30]	In this work, a micro-analytical system for rapid and quantitative analysis by inhibition immunoassay was presented and applied to the detection of folic acid [30]. Considering the system, 8 polymer microchannels of 65-nL volume each and consisting of microelectrodes were embedded in a cartridge so that they could be run simultaneously [30].
Eteshola and Balberg [31]	In this work, fluorescent reactions of a heterogeneous sandwich enzyme-linked immunoassay in an all-PDMS [poly (dimethylsiloxane)] microfluidic device were detected using a cooled charge coupled device (CCD) camera interfaced with an epifluorescence microscope [31]. Eteshola and Balberg demonstrated the extension of this hybrid integrated method to on-chip imaging and quantification of light emission from a biochemical immunoassay in PDMS chip [31].

B. PCR Microfluidics Technology

The miniaturization of biological and chemical analytical devices by micro-electro-mechanical-systems technology represents a great leap in such fields as medical diagnostics, microbial detection and other bio-analysis [32]. Among many miniaturized analytical devices, the polymerase chain reaction (PCR) microchip/microdevices are studied extensively [32]. There are many studies on great aspects of on-chip micromachining (fabrication, bonding and sealing), choice of substrate materials, surface chemistry and architecture of reaction vessel, handling of necessary sample fluid, controlling of three or two-step temperature thermocycling, detection of amplified nucleic acid products, integration with other analytical functional units such as sample preparation, capillary electrophoresis, DNA microarray hybridization, etc. [32]. Some important reports on PCR microfludics technology are presented in Table 2.

Table 2. Some reports on PCR microfluidics technology

Authors	Details
Marcus et al. [33]	Microfluidic single-cell mRNA isolation and analysis was proposed in this work [33]. Marcus et al. demonstrated single-cell mRNA isolation and cDNA synthesis, undertook quantitative calibrations for each step in the process, and determined gene expression in individual cells [33]. Marcus et al. noted that the methods contributed to the setting for highly parallel single-cell gene expression studies [33].
Ottesen et al .[34]	Ottesen et al. used microfluidic digital PCR to amplify and analyze multiple, different genes obtained from single bacterial cells harvested from nature [34].
Gonzalez et al. [35]	In this work, interaction of quantitative PCR components with polymeric surfaces was studied [35]. Gonzalez et al. mentioned that chemical compatibility of polymeric capillaries with DNA dyes should be carefully considered for the design of quantitative microfluidic devices [35].

Authors	Details
Lien et al. [36]	Purification and enrichment of virus samples utilizing magnetic beads on a microfluidic system was reported in this work. Lien et al. reported a new microfluidic system with three integrated functional devices for pumping, mixing and dividing of bio-samples by utilizing micro-electro-mechanical-systems technology [36]. The success in experiment with dengue virus was presented in this report [36].
Abruzzo et al. [37]	Abruzzo et al. reported on identification and validation of biomarkers of IgV(H) mutation status in chronic lymphocytic leukemia using microfluidics quantitative real-time PCR technology [37]. Results of experiments were reported [37].
Beer et al. [38]	Beer et al. reported the first lab-on-chip system for picoliter droplet generation and PCR amplification with real-time fluorescence detection [38]. Beer et al. quoted for new PCR in isolated droplets at volumes 106 times smaller than commercial real-time PCR tools [38].
Cho et al. [39]	Cho et al. reported a completely integrated, germ-specific DNA extraction device utilizing centrifugal microfluidics on a polymer based couple device platform [39]. Cho et al. demonstrated that this new centrifugal microfluidics platform enabled a full integration of complex biological reactions that required multi-step fluidic control [39].
Chang et al. [40]	Chang et al. reported an integrated microfluidic chip for PCR applied usages utilizing digital microfluidic chip technology [40]. Chang et al. concluded that the developed chips completely demonstrated the feasibility of Lab-on-a-Chip by utilizing electro-wetting-on-dielectric-based digital microfluidics [40].
Li et al. [41]	Li et al. used gold nanoparticles to develop a simple microfluidics-based bioassay that is able to recognize and detect specific DNA sequences via conformational change-induced fluorescence quenching [41]. Li et al. said that this method could be regarded as one-pot reaction and this method holds great promise for clinical diagnostics [41].
Marentis et al. [42]	Marentis et al. presented a piezoelectric microfluidic minisonicator and tested its characteristic [42]. Marentis et al.said that usefulness of the minisonicator over macrofluidic implementations included a small sample volume (2.5 microL), reduced energy consumption, and compatibility with other microfluidic blocks [42].
Pilarski et al. [43]	Sensitive detection using microfluidics technology of single cell PCR products from high and low abundance IgH VDJ templates in multiple myeloma was reported in this work [43]. Pilarski et al. revealed use of capillary electrophoresis to detect PCR product amplified from either transcripts (high abundance template) or genomic DNA (low abundance template) encoding clonotypic immunoglobulin heavy chain VDJ of plasma cells from patients with multiple myeloma [43].
Peytavi et al. [44]	Peytavi et al. developed a microfluidic flow cell consisting of a network of chambers and channels molded into a polydimethylsiloxane substrate [44]. Peytavi et al. concluded that this removable microfluidic system for performing microarray hybridization on glass slides was useful for molecular diagnostics and gene profiling [44].
Healy et al. [45]	Healy et al. reported on microbial DNA typing by automated repetitive-sequence-based PCR, repetitive sequence-based PCR (rep-PCR) method, for bacterial strain typing. [45].

C. Microfluidic Electrochemiluminescence Technology

Electrochemiluminescence or electrogenerated chemiluminescence is a form of chemiluminescence in which the light-emitting chemiluminescent reaction is preceded by an electrochemical reaction [46 - 47]. The usefulness of chemiluminescence is retained, but the electrochemical reaction allows the time and position of the light-emitting reaction to be controlled. Microfluidic electrochemiluminescence technology is an advent of electrochemiluminescence technology. Some important reports on microfluidic electrochemiluminescence technology are presented in Table 3.

Table 3. Some reports on microfluidic electrochemiluminescence technology

Authors	Details
Yin et al. [48]	A tertiary amine derivative, 2-(2-aminoethyl)-1-methylpyrrolidine (AEMP) was completely developed as electrochemiluminescence probe within microfluidic chip using electrochemiluminescence detection [48]
Zaytseva et al. [49]	The development of a microfluidic biosensor with fluorescence detection for the rapid, sensitive, and serotype-specific detection of dengue virus was presented by Zaytseva et al. [49]. In this work, the biosensor chip consisted of poly(dimethylsiloxane) substrate with fabricated microchannels and a glass substrate used to seal the microchannels [49].

PRINCIPLE AND APPLIED USAGE OF NANOFLUIDICS

Similar to microfludics, nanofluidics is an important basic technology for producing a diagnostic device. The control of nanovolume of fluid is used instead of microvolume. There are two most commonly used methods for nanostructuring and nanochannel fabrication in nanofluidics [50]. Top-down methods are based on patterning on a large scale while reducing the lateral dimensions to the nanoscale [50]. Bottom-up methods arrange atoms and molecules in nanostructures [50]. However, the advent in nanofludics is less than that of microfluidics. Some important reports on nanofluidic devices are presented in Table 4.

Table 4. Some important reports on nanofluidic devices

Authors	Details
Parikesit et al. [51]	Parikesit et al. presented the analysis of electroosmotic flow in a branched-turn nanofluidic device, which was developed for detection and sorting of single molecules [51]. Parikesit et al. revealed that the deviation between determined and simulated data could be explained by the determined Brownian motion of the tracer-particles, which was not integrated in the simulation [51].
Wang et al. [52]	Wang et al .developed a highly efficient microfluidic sample preconcentration device based on the electrokinetic trapping mechanism enabled by nanofluidic filters [52]. In this work, the device, fabricated by standard photolithography and etching techniques, generates a widened space charge region within a microchannel, which was used to both collect and trap the molecules efficiently [52].

Authors	Details
Riehn et al. [53]	Riehn et al .performed restriction mapping of DNA molecules using restriction endonucleases in nanochannels with diameters of 100-200 nm [53]. Restriction mapping in a nanofluidic device was reported in this work [53].
Riehn and Austin [54]	Riehn and Austin described a method for wetting micro- and nanofluidic devices with hydro or any other pure liquid [54]. In this work, Riehn and Austin demonstrated wetting of micro- and nanostructures in a fused-silica device with only a single inlet [54].
van der Heyden [55]	Electrokinetic energy conversion efficiency in nanofluidic channels was studied [55]. Van der Heyden theoretically assessd the prospect of using electrokinetic phenomena to convert hydrostatic energy to electrical power and reported that an electrokinetic energy conversion device could get a maximum efficiency of 12% for simple monovalent ions in aqueous solution [55].
Karnik et al. [56]	Rectification of ionic present in a nanofluidic diode was studied [56]. Karnik et al. demonstrated rectification of ionic transport in a nanofluidic diode fabricated by introducing a surface charge discontinuity in a nanofluidic channel in this work [56].

NANOBIOSENSOR

As previously mentioned in the early part of this chapter, the biochip is an important diagnostic technology. Nanobiosensor is a new hope in diagnostic technology. Various types of transducer and bioreceptor molecules are presently used for nanosensor and nanobiosensor fabrication [57 - 59]. The details of nanotechnology-based biochips are also presented in the early part of this chapter.

Table 5. Some important reports on nanobiosensor

Authors	Details
Song et al. [60]	In this work, the intracellular measurement of cytochrome c using an optical nanobiosensor was demonstrated [60]. Song et al. said that the combination of the nanobiosensor with the enzyme-linked immunosorbent assay improved the detection sensitivity of the nanobiosensor due to enzymatic amplification [60].
Kasili and Vo- Dinh [61]	In this work, Kasili and Vo-Dinh used nanobiosensors to monitor a molecular signaling process following photodynamic-therapy-induced apoptosis in breast cancer cells [61]. Kasili and Vo-Dinh demonstrated the minutely invasive capability of optical nanobiosensors to measure important signaling molecules and events in pathways at the single cell level in this work [61].
Tripathi et al. [62]	Nanobiosensor design utilizing a periplasmic *E. coli* receptor protein immobilized within Au/polycarbonate nanopores was reported [62].
Risveden et al. [63]	In this work, a new type of bioelectronic region ion-sensitive field effect transistor nanosensor was constructed and demonstrated on two different sensor chips that could measure glucose with good linearity [63]

References

[1] Fodinger M, Sunder-Plassmann G, Wagner OF. Potentials in molecular diagnosis. *Wien Klin Wochenschr.* 1999;111:315-9.

[2] Jain KK. Applied usages of biochips: from diagnostics to personalized medicine.*Curr. Opin. Medicine Discov. Devel.* 2004;7:285-9.

[3] Robinson BH, Seeman NC. The design of a biochip: a self-assembling molecular-scale memory device. *Protein Eng.* 1987;1:295-300.

[4] Odde DJ, Renn MJ. Laser-guided direct writing for applied usages in biotechnology. *Potentials Biotechnol.* 1999;17:385-9.

[5] Kim DS, Lee SH, Ahn CH, Lee JY, Kwon TH. Disposable integrated microfluidic biochip for blood typing by plastic microinjection moulding. *Lab. Chip.* 2006;6:794-802.

[6] Soper SA, Hashimoto M, Situma C, Murphy MC, McCarley RL, Cheng YW, Barany F. Fabrication of DNA microarrays onto polymer substrates using UV modification protocols with integration into microfluidic platforms for the sensing of low-abundant DNA point mutations. *Methods.* 2005;37:103-13.

[7] Besselink GA, Vulto P, Lammertink RG, Schlautmann S, van den Berg A, Olthuis W, Engbers GH, Schasfoort RB. Electroosmotic guiding of sample flows in a laminar flow chamber. *Electrophoresis.* 2004;25:3705-11.

[8] Soo Ko J, Yoon HC, Yang H, Pyo HB, Hyo Chung K, Jin Kim S, Tae Kim Y. A polymer-based microfluidic device for immunosensing biochips. *Lab. Chip.* 2003;3:106-13.

[9] Ionescu RE, Marks RS, Gheber LA. Manufacturing of nanochannels with controlled dimensions using protease nanolithography. *Nano Lett.* 2005;5:821-7.

[10] Albers J, Grunwald T, Nebling E, Piechotta G, Hintsche R. Electrical biochip technology--a tool for microarrays and continuous monitoring. *Anal. Bioanal. Chem.* 2003;377:521-7.

[11] Cui HF, Ye JS, Chen Y, Chong SC, Sheu FS. Microelectrode array biochip: tool for in vitro medicine screening based on the detection of a medicine effect on dopamine release from PC12 cells. *Anal Chem.* 2006;78:6347-55.

[12] Takahashi Y, Nagata T, Nakayama T, Ishii Y, Ishikawa K, Asai S. GeneChip system from a bioinformatical point of view. *Nippon Yakurigaku Zasshi.* 2002;120:73-84.

[13] Zhu T. Global analysis of gene expression using GeneChip microarrays. *Curr. Opin. Plant Biol.* 2003;6:418-25.

[14] Ragoussis J, Elvidge G. Affymetrix GeneChip system: moving from research to the clinic. *Expert Rev. Mol. Diagn.* 2006;6:145-52.

[15] Jane KK. Applied usages of AmpliChip CYP450. *Mol. Diagn.* 2005;9:119-27.

[16] de Leon J, Susce MT, Murray-Carmichael E. The AmpliChip CYP450 genotyping test: Integrating a new clinical tool. *Mol. Diagn. Ther.* 2006;10:135-51.

[17] de Leon J. AmpliChip CYP450 test: personalized medicine has arrived in psychiatry. *Expert Rev. Mol. Diagn.* 2006;6:277-86.

[18] Dupuy AM, Lehmann S, Cristol JP. Protein biochip systems for the clinical laboratory. *Clin. Chem. Lab. Med.* 2005;43:1291-302.

[19] Molloy RM, Mc Connell RI, Lamont JV, FitzGerald SP. Automation of biochip array technology for quality results. *Clin. Chem. Lab. Med.* 2005;43:1303-13.

[20] Haglmuller J, Rauter H, Bauer G, Pittner F, Schalkhammer T. Resonant nano-cluster devices. *IEE Proc. Nanobiotechnol.* 2005;152:53-63

[21] Sokuler M, Gheber LA. Resonant nano-cluster devices. *IEE Proc. Nanobiotechnol.* 2005;152:53-63.

[22] Degenhart GH, Dordi B, Schonherr H, Vancso GJ. Micro- and nanofabrication of robust reactive arrays based on the covalent coupling of dendrimers to activated monolayers. *Langmuir.* 2004;20:6216-24.

[23] Haeberle S, Zengerle R. Microfluidic platforms for lab-on-a-chip applied usages. *Lab. Chip.* 2007 Sep;7(9):1094-110.

[24] Lee LJ, Yang ST, Lai S, Bai Y, Huang WC, Juang YJ. Microfluidic enzyme-linked immunosorbent assay technology. *Adv. Clin. Chem.* 2006;42:255-95.

[25] Lai S, Wang S, Luo J, Lee LJ, Yang ST, Madou MJ. Design of a compact disk-like microfluidic platform for enzyme-linked immunosorbent assay. *Anal. Chem.* 2004 Apr 1;76(7):1832-7.

[26] Morier P, Vollet C, Michel PE, Reymond F, Rossier JS. Gravity-induced convective flow in microfluidic systems: electrochemical characterization and applied usage to enzyme-linked immunosorbent assay tests. *Electrophoresis.* 2004 Nov;25(21-22):3761-8.

[27] Bai Y, Koh CG, Boreman M, Juang YJ, Tang IC, Lee LJ, Yang ST. Surface modification for enhancing antibody binding on polymer-based microfluidic device for enzyme-linked immunosorbent assay. *Langmuir.* 2006 Oct 24;22(22):9458-67.

[28] Moorthy J, Mensing GA, Kim D, Mohanty S, Eddington DT, Tepp WH, Johnson EA, Beebe DJ. Microfluidic tectonics platform: A colorimetric, disposable botulinum toxin enzyme-linked immunosorbent assay system. *Electrophoresis.* 2004 Jun;25(10-11):1705-13.

[29] Herrmann M, Veres T, Tabrizian M. Enzymatically-generated fluorescent detection in micro-channels with internal magnetic mixing for the development of parallel microfluidic ELISA. *Lab. Chip.* 2006 Apr;6(4):555-60.

[30] Hoegger D, Morier P, Vollet C, Heini D, Reymond F, Rossier JS. Disposable microfluidic ELISA for the rapid determination of folic acid content in food products. *Anal. Bioanal. Chem.* 2007 Jan;387(1):267-75.

[31] Eteshola E, Balberg M. Microfluidic ELISA: on-chip fluorescence imaging. *Biomed. Microdevices.* 2004 Mar;6(1):7-9.

[32] Zhang C, Xu J, Ma W, Zheng W. PCR microfluidic devices for DNA amplification. *Biotechnol. Adv.* 2006 May-Jun;24(3):243-84.

[33] Marcus JS, Anderson WF, Quake SR. Microfluidic single-cell mRNA isolation and analysis. *Anal. Chem.* 2006 May 1;78(9):3084-9.

[34] Ottesen EA, Hong JW, Quake SR, Leadbetter JR. Microfluidic digital PCR enables multigene analysis of indivicouple environmental bacteria. *Science.* 2006 Dec 1;314(5804):1464-7.

[35] Gonzalez A, Grimes R, Walsh EJ, Dalton T, Davies M. Interaction of quantitative PCR components with polymeric surfaces. *Biomed. Microdevices.* 2007 Apr;9(2):261-6.

[36] Lien KY, Lin JL, Liu CY, Lei HY, Lee GB. Purification and enrichment of virus samples utilizing magnetic beads on a microfluidic system. *Lab. Chip.* 2007 Jul;7(7):868-75.

[37] Abruzzo LV, Barron LL, Anderson K, Newman RJ, Wierda WG, O'brien S, Ferrajoli A, Luthra M, Talwalkar S, Luthra R, Jones D, Keating MJ, Coombes KR. Identification and validation of biomarkers of IgV(H) mutation status in chronic lymphocytic leukemia using microfluidics quantitative real-time polymerase chain reaction technology. *J. Mol. Diagn.* 2007 Sep;9(4):546-55.

[38] Beer NR, Hindson BJ, Wheeler EK, Hall SB, Rose KA, Kennedy IM, Colston BW. On-Chip, Real-Time, Single-Copy Polymerase Chain Reaction in Picoliter Droplets. *Anal. Chem.* 2007 Oct 11;

[39] Cho YK, Lee JG, Park JM, Lee BS, Lee Y, Ko C. One-step germ specific DNA extraction from whole blood on a centrifugal microfluidic device. *Lab. Chip.* 2007 May;7(5):565-73.

[40] Chang YH, Lee GB, Huang FC, Chen YY, Lin JL. Integrated polymerase chain reaction chips utilizing digital microfluidics. *Biomed. Microdevices.* 2006 Sep;8(3):215-25.

[41] Li YT, Liu HS, Lin HP, Chen SH. Gold nanoparticles for microfluidics-based biosensing of PCR products by hybridization-induced fluorescence quenching. *Electrophoresis.* 2005 Dec;26(24):4743-50.

[42] entis TC, Kusler B, Yaralioglu GG, Liu S, Haeggstrom EO, Khuri-Yakub BT. Microfluidic sonicator for real-time disruption of eukaryotic cells and bacterial spores for DNA analysis. *Ultrasound Med. Biol.* 2005 Sep;31(9):1265-77.

[43] Pilarski LM, Lauzon J, Strachan E, Adamia S, Atrazhev A, Belch AR, Backhouse CJ. Sensitive detection using microfluidics technology of single cell PCR products from high and low abundance IgH VDJ templates in multiple myeloma. *J. Immunol. Methods.* 2005 Oct 20;305(1):94-105.

[44] Peytavi R, Raymond FR, Gagne D, Picard FJ, Jia G, Zoval J, Madou M, Boissinot K, Boissinot M, Bissonnette L, Ouellette M, Bergeron MG. Microfluidic device for rapid (<15 min) automated microarray hybridization. *Clin. Chem.* 2005 Oct;51(10):1836-44.

[45] Healy M, Huong J, Bittner T, Lising M, Frye S, Raza S, Schrock R, Manry J, Renwick A, Nieto R, Woods C, Versalovic J, Lupski JR. Microbial DNA typing by automated repetitive-sequence-based PCR. *J. Clin. Microbiol.* 2005 Jan;43(1):199-207.

[46] Kissinger PT. Electrochemical detection in bioanalysis. *J. Pharm. Biomed. Anal.* 1996;14:871-80.

[47] Richter MM. Electrochemiluminescence (ECL). *Chem. Rev.* 2004;104:3003-36.

[48] Yin XB, Du Y, Yang X, Wang E. Microfluidic chip with electrochemiluminescence detection using 2-(2-aminoethyl)-1-methylpyrrolidine tagging. *J. Chromatogr.* A. 2005 Oct 14;1091(1-2):158-62.

[49] Zaytseva NV, Montagna RA, Baeumner AJ. Microfluidic biosensor for the serotype-specific detection of dengue virus RNA. *Anal. Chem.* 2005 Dec 1;77(23):7520-7.

[50] Mijatovic D, Eijkel JC, van den Berg A. Technologies for nanofluidic systems: top-down vs. bottom-up--a review. *Lab. Chip.* 2005 May;5(5):492-500.

[51] Parikesit GO, Markesteijn AP, Kutchoukov VG, Piciu O, Bossche A, Westerweel J, Garini Y, Young IT. Electroosmotic flow analysis of a branched U-turn nanofluidic device. *Lab. Chip.* 2005 Oct;5(10):1067-74.

[52] Wang YC, Stevens AL, Han J. Million-fold preconcentration of proteins and peptides by nanofluidic filter. *Anal. Chem.* 2005 Jul 15;77(14):4293-9.

[53] Riehn R, Lu M, Wang YM, Lim SF, Cox EC, Austin RH. Restriction mapping in nanofluidic devices. *Proc. Natl .Acad. Sci .U S A.* 2005 Jul 19;102(29):10012-6.

[54] Riehn R, Austin RH. Wetting micro- and nanofluidic devices using supercritical hydro. *Anal. Chem.* 2006 Aug 15;78(16):5933-4.

[55] van der Heyden FH, Bonthuis DJ, Stein D, Meyer C, Dekker C. Electrokinetic energy conversion efficiency in nanofluidic channels. *Nano Lett.* 2006 Oct;6(10):2232-7.

[56] Karnik R, Duan C, Castelino K, Daiguji H, Majumdar A. Rectification of ionic present in a nanofluidic diode. *Nano Lett.* 2007 Mar;7(3):547-51.

[57] Cullum BM, Vo-Dinh T. The development of optical nanosensors for biological measurements. *Potentials Biotechnol.* 2000 Sep;18(9):388-93.

[58] Cui D, Gao H. Advance and prospect of bionanomaterials. *Biotechnol. Prog.* 2003 May-Jun;19(3):683-92.

[59] Cui D, Gao H. Advance and prospect of bionanomaterials. *Biotechnol. Prog.* 2005 Mar-Apr;21(2):650.

[60] Song JM, Kasili PM, Griffin GD, Vo-Dinh T. Detection of cytochrome C in a single cell using an optical nanobiosensor. *Anal Chem.* 2004 May 1;76(9):2591-4.

[61] Kasili PM, Vo-Dinh T. Optical nanobiosensor for monitoring an apoptotic signaling process in a single living cell following photodynamic therapy. *J. Nanosci. Nanotechnol.* 2005 Dec;5(12):2057-62.

[62] Tripathi A, Wang J, Luck LA, Suni II. Nanobiosensor design utilizing a periplasmic E. coli receptor protein immobilized within Au/polycarbonate nanopores. *Anal. Chem.* 2007 Feb 1;79(3):1266-70.

[63] Risveden K, Ponten JF, Calander N, Willander M, Danielsson B. The region ion sensitive field effect transistor, a new bioelectronic nanosensor. *Biosens Bioelectron.* 2007 Jun 15;22(12):3105-12.

Chapter 5

NANOBIODEVICE AND NANOPROBE

INTRODUCTION TO MEDICAL TOOL DEVELOPMENT AND TECHNOLOGY [1 - 18]

Similar to other sciences, medicine arose from observation in its classical period. However, the limitation of human motor and sensory organs led to a limitation in observing and performing medical activities. Finding a new helper was a common focus of medical research at that time. The intervention of many medical tools was the solution for the described problem. Time passed; many new tools were invented. There was a new need to find controls for the medical tools. This is the starting point of medical toolation technology. At present, a multiple appearance of business systems, clinical work systems, tool engineering systems, information systems, infrastructure systems, and management systems interact to make the modern healthcare facility work [19]. The key to designing for such a system is systems engineering, a skill often little appreciated among clinical biomedical engineers [19]. The principal activities of clinical biomedical engineering include eliciting, understanding, negotiating, describing, validating, and managing system requirements [19].

Medical tool technology is an important focus of biomedical engineering. Medical tool technology has a wide scope covering designing, producing, operating, fixing, and controlling of medical tools. Medical tool technologists play important roles in these processes. The important details for each described process will be hereby discussed briefly.

A. Designing

Designing is a necessarily important process in medical tool technology [3]. Design determines the success of medical tool development. Designing should concern the material and method of producing the tool. Expose the design team to the user environment, followed by question-and-answer periods with users while still in the user environment [3]. Place biomedical engineers in the leading teaching institutions where they will have day-to-day exposure to the use of products similar to the one being designed. Bring biomedical engineers back into the company as part of the design team [3]. The team should be told to consider what the actual desired role of the planned medical tool is. The designing should focus on producing a tool that can comply with the desire for keeping costs low. In addition, the developed tool should not cause environmental problems.

B. Production

Production is the step that has to be based on the direction of medical tool engineers. Production should be performed in a specific workshop under a standardized condition. At present production of medical tools can be in-house production or industrial production. In-house production is a process to serve the local need. The in-house production usually delivers a small medical tool that can be used in a local setting, while industrial production takes important roles for large medical tools and would usually be in the form of motion tool manufacture.

Production of a medical tool can be divided into two important parts. The first part is the role for electronic and electrical engineering which usually involves the sensor, data processing, and operating unit in the medical tool. The other part plays an important role in determining the external appearance of the machine, including its strength.

C. Operating

This is the step involving real usage in medical practice. The operating is usually carried out by a physician, nurse or medical technologist. This step can be successful if the users strictly practice according to the operating manual guidelines.

D. Repair

When the medical tool does not remain in good condition, repair is necessary. The principle of repair is trying to resolve the problem if possible. In case repair is not possible or not expected to be effective, finding a new medical tool is suggested.

E. Control

Control is an important step. Quality control of the whole process is necessary. The details of a quality management system for medical devices will be discussed in the next section.

QUALITY MANAGEMENT SYSTEM FOR MEDICAL DEVICES [20 – 26]

A. Introduction

Quality management is the new potential in medicine at present. There have been many quality systems introduced to the medical society within the past few years. The ISO series are a good example of an international quality system that can be applied to medicine (Table 1).

Table 1. Example of important ISO series which are commonly used in medicine

Series	Comments
ISO9000	Key for customer satisfaction
ISO14000	Key for environmental control
ISO17025	Key for laboratory process
ISO15189	Key for medical laboratory process

Applying the principle of three phases of service (Figure 1), the ISO series for all steps are necessary. The consideration for production is the good product, for the process it is the good working method, and for practice it is good participation.

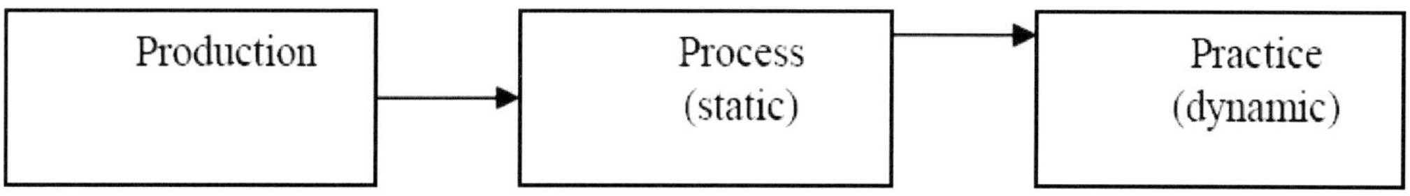

Figure 1. The principle of three phases of service.

For medical devices, the production is important. The quality of the medical device is necessary since the device can directly and indirectly impact the user and patients. A good example of direct effect is the allergy to latex gloves, and of indirect effect the unreliable laboratory result due to an imprecise analyzer. Hence, the quality system for quality control of medical devices is useful. Recently, the new ISO series, ISO 13485, has been launched and is commonly used.

B. ISO13485

ISO13485 is the new ISO series used in the production phase. ISO 13485 certification attests that the quality assurance system of the medical device company complies with regulatory requirements [1]. Presently, ISO 13485 is required by many countries, including European nations, Canada, Taiwan and Australia, for regulatory approval of medical devices. Basically, GMP (Good Manufacturing Practice) has a strong correlation with ISO 13485 [2 – 3].

Presently there are three published standards to consider for registration: ISO13485:1996, ISO13488:1996 and ISO13485:2003. ISO13485:1996 and ISO13488:1996 build on ISO9001:1994 and ISO9002:1994 and replace EN46001:1996 and EN46002:1996 respectively [1]. ISO13485:2003, published in July 2003, is based on the ISO9001:2000 process model approach. This standard will completely replace ISO13485:1996 and ISO13488:1996 by the summer of 2006 [20]. These standards provide a good base model for compliance with the EU CE marking medical device directives, Health Canada CMDCAS, Taiwan Medical Device Regulations and other international requirements [21].

ISO13485:1996, ISO13488:1996 and increasingly ISO13485:2003 are becoming commonly accepted as the international standards to address medical device regulations around the world. Registration to ISO13485 takes place when an accredited third party visits an organisation, assesses the quality management system and, if satisfactory, issues a certificate warranting that the organisation's quality management system reaches the requirements of the Standard [20 – 22]. In addition to assessing quality management systems to the ISO13485: standards, the full range of complementary assessment services such as ISO9001:2000, ISO14001 and ISO18001 should be followed in the production of medical devices.

C. Details of ISO13485 [20 - 23]

There are eight sections for the requirements of ISO13485. Sections 1 to 3 and 7 to 8 cover a multiple appearance of introductory and explanatory topics, which are similar to other ISO series. For section 4 – 6, specific requirements are documented. Section 4 is about the system requirements, section 5 is about management requirements, and section 6 is about resource requirements. The details of the requirements for sections 4 – 6 are presented in Table 2.

Table 2. The details of the requirements for sections 4 – 6 for ISO 13485

Section	Details
Section 4	Establish a quality system for medical devices Document a medical device quality system Support quality
Section 5	Focus on customers Establish a quality policy Perform quality planning Control quality management system Carry out management reviews
Section 6	Bring quality resources Bring quality personnel Bring quality infrastructure Bring quality environment

TYPES OF MEDICAL TOOLS AND NANODEVICES

There are many medical tools at present. The classification of these medical tools is necessary for controlling usage. Overtly the medical tool can be classified into diagnosis device, therapeutic device, and laboratory device. The diagnostic device is a device that acts as a helper for medical diagnostic procedures. Good examples are endoscopes. The therapeutic device is a device that helps in medical therapeutic procedures. Good examples are electrosurgical knife, coagulation, etc. The laboratory device is a medical device used in a medical laboratory and usually under control of a medical technologist and clinical pathologist. Good examples are automated hematology analyzer, automated clinical chemistry analyzer, etc. With the advent of nanotechnology, a new group of medical tool has been launched. This group of medical tools includes so-called nanobiodevice. A nanobiodevice is a medical tool that has a specific small size at the range of nanolevel. These nanobiodevices can be used for diagnosis or therapy. A good property of a nanobiodevice is its small size. Penetration into small parts is feasible, and control of nanovolume substances is also feasible. The details of some important nanobiodevices at present will be discussed later in this chapter.

The recent development of nanobiotechnologies based on the nanofabrication and molecular nanotechnology help diagnose diseases [27 - 28]. The technologies allow us to develop new methods to separate many biomolecules, including DNA, RNA, protein, peptide, polysaccharide, and glycoprotein towards the applied usage in genomics, transcriptomics, proteomics, and glycoproteomics [27]. Nanoparticles, such as gold nanoparticles and quantum dots, can be effectively used in producing DNA-based biodevices but various other nanotechnologies for manipulation at nanoscale as nanobiosensors are also mentioned [29]. This method has the potential to allow controlled placement of particles with subnanometer precision and to allow attachment of the product to fixed contacts for nanodevice fabrication [30]. These technologies will extend the limits of present molecular diagnostics and enable point-of-care diagnosis in the development of personalized medicine [29]. In the longer term, molecular machine systems and medical nanorobots will allow instant germ diagnosis and extermination [30]. It can be said that the advent of the diagnostic nanodevice greatly helps diagnosis in medicine. Here, the details of some important groups of nanodevices are discussed.

A. DNA-Based Nanodevice

As previously mentioned, a DNA-based nanodevice has been developed and is in use at present [31–32]. The specific molecular recognition properties and large aspect ratio of DNA make the molecule a useful template for bottom-up fabrication of nanowires and nanodevices [31]. Fabricating well-defined DNA-templated nanowires requires aligned surface deposition and specific metallization of DNA molecules [31]. DNA localization on surfaces can be obtained by bulk fluid flow or a moving air-hydro interface, and localization efficiency has been improved by surface modifications that designed DNA-substrate interaction [31]. DNA-templated nanowires can also be constructed from gold, silver, copper, palladium, and platinum, and template modifications have allowed the bottom-up construction of a simple electronic nanodevice [31]. For example, Stevenson et al. studied covalent attachment of gold nanoparticles to DNA templates [33]. In this work, functionalized gold nanoparticles were covalently bound to internal, modified sites on double-stranded DNA [33]. More examples are shown in Table 3.

Table 3. Some interesting reports on DNA–based nanodevices

Authors	Details
Li et al. [34]	Li et al. reported, for the first time, that DNA destabilization and conformational transition induced by single-walled carbon nanotubes (SWNTs) are sequence-dependent [34]. Contrasting changes for SWNTs binding to poly[dGdC]:poly[dGdC] and poly[dAdT]:poly[dAdT] were observed by Li et al. [34]. Li et al. provided direct evidence that SWNTs induced DNA B-A transition in solution and they bound to the DNA major groove with GC preference [34].
Bjork et al. [35]	Bjork et al. reported on soft lithographic printing of patterns of stretched DNA and DNA/electronic polymer wires by surface-energy modification and transfer [35]. As an applied usage example, transfer printing of DNA decorated with a semiconducting conjugated polyelectrolyte could be demonstrated in this study [35]. In addition, the resulting patterned localization of stretched DNA could be utilized for functional nanodevice structures, and for biological applied usages [35].
Mao et al. [36]	Mao et al. reported a swift-switching DNA nanodevice that was amplified by an alternating electric field [36]. Mao et al. reported that the device, self-integrated from folded four-stranded DNA motifs, could switch between closed and open states in smart response to pH stimulus, of which the closed state forms a nanometer-height which was impermeable to small molecules [36].
Reuter et al. [37]	In this work, fluorescence resonance energy transfer and fluorescence correlation spectroscopy experiments were used to explore the kinetics of the release process in detail, and to this end the influence of concentration variations and also of temperature were detected [37]. Reuter et al. said that the results from the experiments and a numerical simulation supported the view that the DNA effector strand first binded to the aptamer device followed by the displacement of the protein [37].
Singer and Smith [38]	In this work, Singer and Smith used DNA Y-junctions as fluorescent scaffolds for EcoRII methyltransferase-thioredoxin (M.EcoRII-Trx) fusion proteins [38]. Covalent links between the DNA scaffold and the methyltransferase were formed at preselected sites on the scaffold consisting of 5FdC [38]. Singer and Smith suggested that this approach might serve as an addition to immunohistochemical methods in tumor classification and probe cell surface receptor architecture and function [39].

DNA nanobiochip is the specific DNA-based bionanodevice [39]. Basically, microelectronic arrays have been developed for DNA hybridization analysis of point mutations, single nucleotide polymorphisms, short tandem repeats and gene expression [40]. In addition to a multiple appearance of molecular biology and genomic research applied usages, such devices will also be used for infectious disease detection, genetic and cancer diagnostics, and pharmacogenomic applied usages [40]. DNA nanochip is an important diagnostic nanodevice that can serve the previously-mentioned purposes. Nanogen NanoChip is the best example of a DNA nanochip. The NanoChip platforms employ hybridization-based technology, using fluorescent detection and electronic control of the target or probe, to obtain clear genotype signal relative to background, and increased flexibility relative to similar chip-based SNP genotyping platforms [41]. The NanoChip Molecular Biology Workstation allows the active deposition and concentration of charged biotinylated molecules on designated test sites [42]. The DNA at each pad is then hybridized with specific oligonucleotide probes, complementary to normal or mutant sequences, that are labeled with Cy3 or Cy5 dyes, respectively [42]. Single-nucleotide polymorphism and mutation identification by the Nanogen NanoChip are probable [42]. In 2007, Lee et al. reported on validation of a CYP2D6 genotyping panel on the NanoChip Molecular Biology Workstation [43]. Lee et al. compared CYP2D6 genotypes and resolved samples with genotypic discrepancies with the Jurilab, a long-range PCR assay used to assess sequence structures 3' of the CYP2D7 and CYP2D6 coding regions, CYP2D6 duplication/deletion assay or with traditional DNA sequencing [43]. Lee et al. found that this validation study of the NanoChip electronic microarray-based CYP2D6 genotyping assay revealed a good coefficient of variance and good concordance with the Pyrosequencing method and a warrantatory sequencing method [43]. In 2005, Borsting et al. described a SNP typing protocol developed for the NanoChip electronic microarray [44]. Borsting et al. reported that base stacking between the short reporter and the longer stabilizer oligo stabilized the binding of a matching reporter, whereas the binding of a reporter carrying a mismatch in the single-nucleotide polymorphism position would be relatively weak [44]. Borsting et al. also mentioned that thermal stringency was applied to the NanoChip array according to a reader protocol generated by the user and the fluorescent label on the matching reporter was detected [44]. In 2004, Borsting et al. tested the single-nucleotide polymorphism typing protocol developed for the NanoChip electronic microarray by analyzing the four Y chromosome loci SRY1532, SRY8299, TAT, and 92R7 [45]. In a comparative study, Borsting et al. typed the 5 Y chromosome loci M173, 92R7, P25, SRY1532, and M9 in 400 males using the NanoChip single-nucleotide polymorphism typing protocol and the SNaPshot kit [45]. Borsting et al. said that concording results were obtained for all samples demonstrating the accuracy of the NanoChip single-nucleotide polymorphism typing protocol [45]. Evans and Lee-Tataseo reported a determination of the factor V Leiden single-nucleotide polymorphism in a commercial clinical laboratory by use of NanoChip microelectronic array technology [46]. In their work, primers specific for factor V were treated, and polymerized chain reaction was performed and samples were analyzed using the NanoChip Molecular Biology Workstation with fluorescently labeled reporters for wild-type and single-nucleotide polymorphism sequences [46]. According to this work, Evans and Lee-Tataseo concluded that the NanoChip microelectronic chip array technology was an accurate and convenient method for Factor V Leiden screening of research samples in a clinical laboratory environment [46]. Schrijver et al. performed a similar work [47]. In this work, they compared the NanoChip and other methods for accuracy, time to results, throughput, and interpretation [47]. Schrijver et al. found that almost all from individual amplicons analyzed on the NanoChip were in complete agreement with the other assays [47]. Schrijver et al. concluded that the NanoChip assay is reliable and may be especially valuable to laboratories with a large volume of thrombophilia test requests [47].

B. RNA-Based Nanodevice

Precise quantification and quality characterisation of isolated RNAs are basic requirements for their further applied usages in genome-wide microarrays, northern blots, cDNA library preparation and others. In addition to the DNA-based nanodevice, the RNA-based nanodevice can also be seen in routine practice. RNA Nano Assay is the best example. Ricicova and Palkova indicated that RNA analyses using Agilent RNA Nano Assay exhibited great usefulness when compared with those performed on ethidium bromide-stained agarose gel electrophoresis or on a spectrophotometer [48]. In addition, Ricicova and Palkova also found that the existence of impurities including traces of DNA within RNA samples did not influence the concentration measurements [48]. Other interesting reports on RNA-based biodevices are presented in Table 4. In addition to the diagnostic RNA nanodevices, therapeutic RNA nanodevices can also be seen.

Table 4. Some interesting reports on RNA-based nanodevices

Authors	Details
Nakamura et al. [49]	Nakamura et al. succeeded in producing an efficient method for packaging siRNA into the octaarginine-modified multifunctional envelope-type nano device (R8-MEND), and its silencing effect was compared with that of transfection reagent TransIT-TKO [49]. Nakamura et al. noted that R8-MEND can deliver condensed siRNA into cells to produce an efficient and persistent silencing effect with minimum cytotoxicity [49].
Moriguchi et al. [50]	A MEND for use in the delivery of siRNA expression plasmids was described in this work [50]. Moriguchi et al. demonstrated that MEND was a useful gene delivery system for siRNA expression plasmids with less heterogeneity associated with the transfection [50].

C. Peptide-Based Nanodevice

Peptide-based nanodevice is another type of diagnostic nanodevice. This group is based on the peptide. Some interesting reports on peptide-based nanodevices are shown in Table 5.

Table 5. Some interesting reports on DNA-based nanodevices

Authors	Details
Lee et al. [51]	In this work, PEG-conjugated immunodominant peptides for collagen-induced arthritis were treated for oral tolerance induction instead of whole Type II collagen (CII), because a small peptide can be converted to a macromolecule soluble in methylene chloride by the coupling of poly-ethylene glycol (PEG) [51]. According to this work, it was demonstrated that nanoparticles entrapping a PEG-conjugated peptide could be an alternative delivery method for the induction of oral tolerance rather than CII and peptide [51].
Itoh et al. [52]	Itoh et al. reported on the development of a new biomaterial, hydroxyapatite/collagen (HAp/Col) composite for medical use [52].

D. Carbon Nanotube-Based Nanodevice

As mentioned in the previous chapter, the carbon nanotube, a two-dimensional nanopolymer, can be used for producing nanodevices [53 – 56]. Both diagnostic and therapeutic nanodevices can be developed based on carbon nanotubes. As the leading nanodevice candidate, single-walled carbon nanotubes have potential therapeutic applied usages in gene therapy and new medicine delivery [57]. The miniaturization of nanotube devices for biological applied usages requires complete understanding of the nanotube interaction mechanism [58]. Li et al. found that single-walled carbon nanotubes could inhibit DNA duplex association and selectively induce human telomeric i-motif DNA formation by binding to the 5'-end major groove under physiological conditions or even at pH 8.0 [57]. This is of great interest for medicine design and cancer therapy. More details on carbon nanotubes are provided in the chapter on nanopolymer in this book.

One of the well-known forms of these described nanodevices is nanobiosensors [59]. More details on nanobiosensors are given in another chapter in this book.

NANOPROBE

A nanoprobe is a technological nanodevice. It integrates biological probes coupled to a transducer. Nanoprobe technology is an important part of nanodevice technology. The nanoprobe can also be applied the medical use. Diagnosis and therapy based on nanoprobes can be implemented. For diagnosis, a single-cell nanoprobe method to examine gene expression of individual living cells using atomic force microscopy without killing the cells is a good example [60 – 61]. Uehara et al. compared the single-cell nanoprobe method with in situ hybridization (ISH) and reported good single-cell nanoprobe [61]. Vo-Dinh et al. reported on antibody-based nanoprobe for measurement of a fluorescent analyte in a single cell [62]. Vo-Dinh et al. concluded that the nanoprobe employs antibody-based receptors targeted to a fluorescent analyte, benzopyrene tetrol (BPT), a metabolite of the carcinogen benzo[a]pyrene (BaP) and of the BaP-DNA adduct [62]. Vo-Dinh et al. noted that detection of BPT was of great biomedical interest, since this species could serve as a biomarker for monitoring DNA damage due to BaP exposure and for probable precancer diagnosis [62]. In addition to the single-cell nanoprobe method, nanoprobe tagging and transmission electron microscopy of cytoplasmic face-up sheets of cell membrane has also developed as a high-resolution approach to map the interactions of proteins and lipids during cell signaling [63]. Kasili et al. demonstrated, for the first time, the applied usage and utility of a unique optical sensor having a nanoprobe for monitoring the onset of the mitochondrial pathway of apoptosis in a single living cell by detecting enzymatic activities of caspase-9 [64]. By comparing the fluorescence signals from apoptotic cells induced by photodynamic treatment and nonapoptotic cells, Kasili et al. completely detected caspase-9 activity, which indicates the onset of apoptosis in the cells [64]. Magnetic nanoprobe is also proposed. A paramagnetic nanoprobe to detect tumor cell death using magnetic resonance imaging (MRI) was reported by Neves et al. [65]. Neves et al. said that this nanoprobe bound with high affinity and specificity to the phosphatidylserine exposed by dying cells and was demonstrated to allow MRI detection of apoptotic tumor cells in vitro [65]. Won et al. developed a technology called magnetism-based interaction capture (MAGIC) that identifies molecular targets on the basis of induced motion of superparamagnetic nanoparticles inside living cells [66]. Won et al. reported that MAGIC could be used to monitor signal-dependent modification and multiple interactions of proteins [66]. Veiseh et al reported on optical and MRI multifunctional nanoprobe for targeting gliomas [67]. In their work, the nanoprobe was synthesized by coating iron oxide nanoparticles with covalently bound bifunctional poly(ethylene glycol) (PEG) polymer, which were then functionalized with chlorotoxin and the near-infrared fluorescing molecule Cy5.5 [67]. A new PAMAM dendrimer-based nanoprobe for couple magnetic resonance and fluorescence imaging modalities was synthesized by Talanov et al. [68]. This dendrimer-based nanoprobe could express couple modality magnetic resonance and fluorescence imaging [68]. Based on the good properties of nanoprobes for diagnosis, the tracing of disease can be the facet of therapeutic usefulness of nanoprobes.

References

[1] Liebig PS. The use of high technology for health care at home: issues and implications. *Med. Instrum.* 1988 Oct;22(5):222-5.

[2] Slutsky AS. High frequency oscillations: physiologic considerations. *Med Instrum.* 1985 Sep-Oct;19(5):199-202.

[3] Woodring PL. Applying technology to operator requirements in medical equipment design. *Med. Instrum.* 1986 Jan-Feb;20(1):6-7.

[4] Pilot LR. Device development and FDA scientific review. *Med Instrum.* 1983 Sep-Oct;17(5):341-2.

[5] Harris M. Legislation to regulate medical devices. Biomater. *Med. Devices Artif. Organs.* 1975;3(3):261-75.

[6] Lee AS. Review of procedures for investigational device exemptions. *J. Clin. Eng.* 1979 Apr-Jun;4(2):121-9.

[7] Franke KA. Safe and effective medical devices--everyone's responsibility. *Gastroenterol. Nurs.* 1992 Apr;14(5):245-8.

[8] Alder HC. Safe Medical Devices Act: management guidance for hospital compliance with the new FDA requirements. *Hosp. Technol. Ser.* 1993 Oct;12(11):1-27.

[9] Marlowe DE, Phillips PJ. FDA recognition of consensus standards in the premarket notification program. *Biomed. Instrum. Technol.* 1998 May-Jun;32(3):301-4.

[10] Medical device legislation and the FDA panel on review on dental devices. Council on Dental Materials and Devices. *J. Am. Dent. Assoc.* 1977 Feb;94(2):353-8.

[11] Medical device regulations: the Bureau of Medical Devices and the Bureau of Radiological Health. *Radiol. Manage.* 1980 Spring;2(4):25-38.

[12] Monsein LH. The regulation of medical devices: new tools for the FDA. *J. Clin. Eng.* 1977 Jul-Sep;2(3):231-40.

[13] Rich S. Better health care with quality medical devices: FDA on the cutting edge of device technology. *Int. J. Trauma Nurs.* 2002 Jul-Sep;8(3):89-90.

[14] Medical devices--establishment of procedures to make a device a banned device: final rule. *Fed. Regist.* 1979 May 18;44(98):29213-24.

[15] United States government regulation of medical device software: a review. *J. Med. Eng. Technol.* 1990 May-Jun;14(3):111-3.

[16] Medical devices; intent to initiate proceedings to establish effective dates of requirement for premarket approval for 31 class III preamendments devices--FDA. Advance notice of proposed rulemaking. *Fed. Regist.* 1989 Jan 6;54(4):550-2.

[17] Munsey RR. Potentials and events in FDA regulation of medical devices over the last fifty years. *Food Medicine Law J.* 1995;50 Spec:163-77.

[18] Crumpler ES, Rudolph H. FDA software policy and regulation of medical device software. *Food Medicine Law J.* 1997;52(4):511-6.

[19] Carr JJ, McCullough CE. Requirements engineering: the key to designing complex medical systems. *Biomed. Instrum. Technol.* 2000 May-Jun;34(3):185-95.

[20] Donawa M. Update on ISO 13485. *Med. Device Technol.* 2002 Mar;13(2):23-5.

[21] Donawa M. The impact of the new quality management system standards on the CE-marking process. *Med. Device Technol.* 2001 Jan-Feb;12(1):34-9.

[22] Gundarov VP, Kavalerov GI. Development of standardization and certification of the quality of medical devices. *Med. Tekh.* 2001 May-Jun;(3):4-8.

[23] Ward A. ANSI/AAMI/ISO 13485 comes of age. *Biomed. Instrum. Technol.* 2007 Mar-Apr;41(2):145-6.

[24] Mashkov AV, Golikov VA. Certification of medical devices. *Med. Tekh.* 2003 Jan-Feb;(1):51-3.

[25] Lally R. Achieving systematic compliance with the Medical Device Directive. *Med. Device Technol.* 1996 Sep;7(7):28-33.

[26] Donawa M. European medical device regulation: a new era? *Med Device Technol.* 2004 Dec;15(10):30-1.

[27] Amato I.Step by Step to a Nanodevice. *Science.* 1991 Nov 29;254(5036):1302-1303.

[28] Baba Y. Diagnosis of diseases by nanodevice. *Nippon Rinsho.* 2006 Feb;64(2):271-7.\

[29] Jain KK. Nanotechnology in clinical laboratory diagnostics. *Clin. Chim. Acta.* 2005 Aug;358(1-2):37-54.

[30] Freitas RA Jr. The future of nanofabrication and molecular scale devices in nanomedicine. *Stud. Health Technol. Inform.* 2002;80:45-59.

[31] Stoltenberg RM, Woolley AT. DNA-templated nanowire fabrication. *Biomed. Microdevices.* 2004 Jun;6(2):105-11.

[32] Rusk N. Janus face of DNA. *Nat Methods.* 2006 May;3(5):334-5.

[33] Stevenson KA, Muralidharan G, Maya L, Wells JC, Barhen J, Thundat T. Covalent attachment of gold nanoparticles to DNA templates. *J. Nanosci. Nanotechnol.* 2002 Jun-Aug;2(3-4):397-404.

[34] Li X, Peng Y, Qu X. Carbon nanotubes selective destabilization of duplex and triplex DNA and inducing B-A transition in solution. *Nucleic. Acids Res.* 2006 Aug 2;34(13):3670-6.

[35] Bjork P, Holmstrom S, Inganas O. Soft lithographic printing of patterns of stretched DNA and DNA/electronic polymer wires by surface-energy modification and transfer. *Small.* 2006 Aug;2(8-9):1068-74.

[36] Mao Y, Liu D, Wang S, Luo S, Wang W, Yang Y, Ouyang Q, Jiang L. Alternating-electric-field-amplified reversible switching of DNA nanoconsist ofers with pH. Nucleic *Acids Res.* 2007;35(5):e33.

[37] Reuter A, Dittmer WU, Simmel FC. Kinetics of protein-release by an aptamer-based DNA nanodevice. *Eur. Phys. J. E Soft Matter.* 2007 Jan;22(1):33-40.

[38] Singer EM, Smith SS. Nucleoprotein assemblies for cellular biomarker detection. *Nano Lett.* 2006 Jun;6(6):1184-9.

[39] Baby Y. Ultrafast analysis of DNA by microchip and nanochip technology. Tanpakushitsu Kakusan Koso. 2000 Jan;45(1):76-85.

[40] Sosnowski R, Heller MJ, Tu E, Forster AH, Radtkey R.Active microelectronic array system for DNA hybridization, genotyping and pharmacogenomic applied usages. *Psychiatr. Genet.* 2002 Dec;12(4):181-92.

[41] -Kim D, Grody WW, Richards CS. Microelectronic array system for molecular diagnostic genotyping: Nanogen NanoChip 400 and molecular biology workstation. *Expert .Rev. Mol. Diagn.* 2006 May;6(3):287-94.

[42] Ferrari M, Cremonesi L, Bonini P, Foglieni B, Stenirri S. Single-nucleotide polymorphism and mutation identification by the nanogen microelectronic chip technology. *Methods Mol. Med.* 2005;114:93-106.

[43] Lee HK, Lewis LD, Tsongalis GJ, Schur BC, Jannetto PJ, Wong SH, Yeo KT. Validation of a CYP2D6 genotyping panel on the NanoChip Molecular Biology Workstation. *Clin. Chem.* 2007 May;53(5):823-8.

[44] Borsting C, Sanchez JJ, Morling N. SNP typing on the NanoChip electronic microarray. *Methods Mol. Biol.* 2005;297:155-68.

[45] Borsting C, Sanchez JJ, Morling N. Multiplex PCR, amplicon size and hybridization efficiency on the NanoChip electronic microarray. *Int. J. Legal Med.* 2004 Apr;118(2):75-82.

[46] Evans JG, Lee-Tataseo C. Determination of the factor V Leiden single-nucleotide polymorphism in a commercial clinical laboratory by use of NanoChip microelectronic array technology. *Clin. Chem.* 2002 Sep;48(9):1406-11.

[47] Schrijver I, Lay MJ, Zehnder JL. Diagnostic single nucleotide polymorphism analysis of factor V Leiden and prothrombin 20210G > A. A comparison of the Nanogen Eelectronic Microarray with restriction enzyme digestion and the Roche LightCycler. *Am. J. Clin. Pathol.* 2003 Apr;119(4):490-6.

[48] Ricicova M, Palkova Z. Comparative analyses of Saccharomyces cerevisiae RNAs using Agilent RNA 6000 Nano Assay and agarose gel electrophoresis. *FEMS Yeast Res.* 2003 Oct;4(1):119-22.

[49] Nakamura Y, Kogure K, Futaki S, Harashima H. Octaarginine-modified multifunctional envelope-type nano device for siRNA. *J. Control Release.* 2007 Jun 22;119(3):360-7.

[50] Moriguchi R, Kogure K, Akita H, Futaki S, Miyagishi M, Taira K, Harashima H. A multifunctional envelope-type nano device for new gene delivery of siRNA plasmids. *Int. J. Pharm.* 2005 Sep 14;301(1-2):277-85.

[51] Lee WK, Park JY, Jung S, Yang CW, Kim WU, Kim HY, Park JH, Park JS. Preparation and characterization of biodegradable nanoparticles entrapping immunodominant peptide conjugated with PEG for oral tolerance induction. *J. Control Release.* 2005 Jun 20;105(1-2):77-88.

[52] Itoh S, Kikuchi M, Koyama Y, Matumoto HN, Takakuda K, Shinomiya K, Tanaka J. Development of a new biomaterial, hydroxyapatite/collagen (HAp/Col) composite for medical use. *Biomed. Mater Eng.* 2005;15(1-2):29-41.

[53] Sinnott SB. Chemical functionalization of carbon nanotubes. *J. Nanosci. Nanotechnol.* 2002 Apr;2(2):113-23.

[54] Boczkowski J, Lanone S. Potential uses of carbon nanotubes in the medical field: how worried should patients be? *Nanomed.* 2007 Aug;2(4):407-10.

[55] Helland A, Wick P, Koehler A, Schmid K, Som C. Reviewing the environmental and human health knowledge base of carbon nanotubes. *Environ. Health Perspect.* 2007 Aug;115(8):1125-31.

[56] Koyama S, Haniu H, Osaka K, Koyama H, Kuroiwa N, Endo M, Kim YA, Hayashi T. Medical applied usage of carbon-nanotube-filled nanocomposites: the microcatheter. *Small.* 2006 Dec;2(12):1406-11.

[57] Li X, Peng Y, Ren J, Qu X. Carboxyl-modified single-walled carbon nanotubes selectively induce human telomeric i-motif formation. *Proc. Natl. Acad. Sci. U S A.* 2006 Dec 26;103(52):19658-63.

[58] Li X, Peng Y, Qu X. Carbon nanotubes selective destabilization of duplex and triplex DNA and inducing B-A transition in solution. *Nucleic. Acids Res.* 2006 Aug 2;34(13):3670-6.

[59] Cao R, Villalonga R, Fragoso A. Towards nanomedicine with a supramolecular approach: a review. *IEE Proc. Nanobiotechnol.* 2005 Oct;152(5):159-64.

[60] Osterreicher J, Vogt J, Tanner JM, Lehmann D, Spemann D, Reinert T, Butz T, Navratil L, Kuna P. Cell analysis with the new Leipzig high-energy ion nanoprobe. *Radiats Biol. Radioecol.* 2003 Mar-Apr;43(2):223-6.

[61] Uehara H, Kunitomi Y, Ikai A, Osada T. mRNA detection of indivicouple cells with the single cell nanoprobe method compared with in situ hybridization. *J. Nanobiotechnology.* 2007 Oct 10;5(1):7.

[62] Vo-Dinh T, Alarie JP, Cullum BM, Griffin GD. Antibody-based nanoprobe for measurement of a fluorescent analyte in a single cell. *Nat. Biotechnol.* 2000 Jul;18(7):764-7.

[63] Zhang J, Leiderman K, Pfeiffer JR, Wilson BS, Oliver JM, Steinberg SL. Characterizing the topography of membrane receptors and signaling molecules from spatial patterns obtained using nanometer-scale electron-dense probes and electron microscopy. *Micron.* 2006;37(1):14-34.

[64] Kasili PM, Song JM, Vo-Dinh T. Optical sensor for the detection of caspase-9 activity in a single cell. *J. Am. Chem. Soc.* 2004 Mar 10;126(9):2799-806.

[65] Neves AA, Krishnan AS, Kettunen MI, Hu DE, Backer MM, Davletov B, Brindle KM. A paramagnetic nanoprobe to detect tumor cell death using magnetic resonance imaging. *Nano Lett.* 2007 May;7(5):1419-23.

[66] Won J, Kim M, Yi YW, Kim YH, Jung N, Kim TK. A magnetic nanoprobe technology for detecting molecular interactions in live cells. *Science.* 2005 Jul 1;309(5731):121-5.

[67] Veiseh O, Sun C, Gunn J, Kohler N, Gabikian P, Lee D, Bhattarai N, Ellenbogen R, Sze R, Hallahan A, Olson J, Zhang M. Optical and MRI multifunctional nanoprobe for targeting gliomas. *Nano Lett.* 2005 Jun;5(6):1003-8.

[68] Talanov VS, Regino CA, Kobayashi H, Bernardo M, Choyke PL, Brechbiel MW. Dendrimer-based nanoprobe for couple modality magnetic resonance and fluorescence imaging. *Nano Lett.* 2006 Jul;6(7):1459-63.

Chapter 6

NANOTHERAPY

WHY "NANOTHERAPY"? [1 - 5]

Treatment is an important medical activity. Although it is accepted that prevention is better than treatment, therapy for an already existing disorder is necessary. Therapy can be either specific or symptomatic treatment. The main aim of treatment is curative or to decrease the suffering of the patient. Development of treatment has been continuous for centuries. The advent of nanotherapy in medical treatment has become a new hope in medicine.

At present, it seems that there are still many limitations of classical treatment. The specificity of treatment in classical therapy is a focus. It is accepted that classical therapy is limited in site and size of treatment. Access to the right site is the key for the most powerful treatment. It cannot be repeated too often that the best connection is the connection at the actual problematic site. This is the concept of medical treatment. Indeed, researchers have proposed for the concept of the magic bullet since the 1990s [6 – 8]. This concept is the target for pharmacological treatment at present. The best example is the case of chemotherapy. In the classical chemotherapy, lack of good site specificity induces many complications, such as destruction of noncancerous cells [9 – 11]. For size specificity, the best example is the concern about surgical wounds [12 – 14]. Now, the concept of surgery at present is that which causes the least injury. This concept also corresponds to the basic medical principle, "First, do no harm" [15 – 18].

The answer to the query "How can we find the site- and size-specific therapy?" has been searched for years. Luckily, due to the advanced nanobiotechnology, the answer seems to have been reached. The core concept of nanotechnology is the "small". In one sense, small means specific: focus on a limited-scope site and size. In addition, due to the principle of nanomaterial, including surface area effect with desired effect, applied usage of nanomaterials as a new supplement to improve classical therapeutic protocol can be very useful. In this chapter, the details on advanced nanotherapy will be summarized and presented.

NANOPHARMACOLOGY FOR NANOTHERAPY

Nanopharmacology is a very useful medical science. It can be applied to nanotherapy. New specific medicine design can be performed at present [19 – 20]. Nanopharmacology is the hope for treatment of presently untreatable diseases such as cancer and human immunodeficiency virus. Details on nanopharmacology can be seen in another chapter in this book.

NANOSURGERY

Nanosurgery is the new concept in surgical medicine [21 - 22]. As previously mentioned, a small wound is a desired outcome of present surgery. The concept is that a small surgical wound means small blood loss and minimization of intraoperative and postoperative complication. A small wound also decreases the postoperative hospitalization period for the patient. However, actual nanosurgery for human beings has rarely been achieved. There are some advents in eye surgery. Femtosecond laser pulses, emitted from lasers working in the near infrared, based on multiphoton effects allowing both imaging and laser effects to be generated which are in the submicron range and which do not cause collateral damage are available [23]. Some interesting reports on nanosurgery are quoted in Table 1.

Table 1. Some interesting reports on nanosurgery

Authors	Details
Tirlapur and Konig [24]	In this work, Tirlapur and Konig described a highly reproducible femtosecond near-infrared laser-based nanoprocessing technique that could be used both for non-invasive intra-tissue nanodissection of plant cell walls and as selective destruction of a single plastid or part thereof without compromising the viability of the cells [24]. Tirlapur and Konig said that the potential applied usages of this technical advance included developmental biology applied usages, particularly studies addressing spatio-temporal control of ontogenetic events and cell-cell interactions, and gravitational biology applied usages [24].
Firtel et al. [25]	Firtel et al. reported their experience in a nanosurgery experiment. In this work, the internal cell wall structure of the bacterium *Lactobacillus helveticus* was observed in situ in aqueous solution using an atomic force microscope (AFM); then the AFM tip was used not only for imaging but presumably to mechanically remove large patches of the outer cell wall after appropriate chemical treatment, which typically leaves the bacteria alive [25].
Sacconi et al. [26]	Sacconi et al. used near-infrared femtosecond laser pulses for a combination of microscopy and nanosurgery on fluorescently labeled structures within living cells [26]. Sacconi et al. reported that nanosurgery could be performed with submicrometer precision and without visible collateral damage to the cell [26].

Authors	Details
Kohli et al. [27]	Membrane surgery and nanosurgical cell isolation using high-intensity femtosecond laser pulses was reported in this work [27]. Kohli et al. demonstrated the applicability of using ultrashort (femtosecond) laser pulses for performing surgery on live mammalian cells [27].
Colombelli et al. [28]	Colombelli reported on the manipulation of intracellular filaments using a nanosurgery system based on a subnanosecond pulsed UV laser optimized for the localized severing of biological polymers [28]. By inducing artificial catastrophe of selected microtubules, Colombelli et al. performed shrinkage-rate measurements in interphase Ptk-2 cells throughout the entire cell [28]. Colombelli concluded that this new method could help the quantitative investigation of cytoskeleton dynamics in a local confinement [28].
Wang and Halbhuber [29]*	Wang and Halbhuber reported on corneal multiphoton microscopy and intratissue optical nanosurgery by nanojoule femtosecond near-infrared pulsed lasers [29]. In this work, multiphoton microscopy including multiphoton autofluorescence imaging and second-harmonic generation was used as a new diagnostic tool to perform tissue nonlinear optical tomography with submicron resolution [29].
Jeffries et al [30] *	Jeffries et al. described the use of a polarization-shaped optical vortex trap, which exerts less photodamage on the trapped particle than conventional optical tweezers, for carrying out single-cell nanosurgical procedures [30]. Jeffries et al. said that this method was also anticipated to find broad use in the trapping of any nanoparticles that were adversely impacted by high-intensity laser light [30].
Wakida et al. [31]	In this study, 532-nm picosecond and 800-nm femtosecond lasers were used in combination with fluorescently labeled tubulin to further elucidate microtubule depolymerization and the effect lasers might have on the resulting depolymerization [31]. Wakida et al. could demonstrate the ability of both the picosecond and femtosecond lasers to cut individual microtubules in this work [31].
Colombelli et al. [32]	In this work, Colombelli discussed the characteristic of the ablation process in terms of the energy deposited onto the sample and compare our implementation to others recently employed for cellular and subcellular surgery [32]. Above the energy threshold of ionization, Colombelli et al. demonstrated how to get single-cell ablation through the induction of mechanical perturbation and cavitation in living organisms [32].

*report that focuses on nanosurgery in ophthalmology.

Nanoneurosurgery is another interesting emerging nanosurgery [33 – 35]. Many advents in nanodiagnosis can be applied to nanosurgery [33]. Targeted nanotherapy based on nanopharmacology is also useful in nanosurgery [34]. The applied usage of nanobiotechnology for nanosurgery can also be easily described as the development of many nanosurgical tools and nanoprosthetic devices. In addition, nanocoated tools can also be useful for surgery and orthopedics. Although bone is a very diverse tissue providing different functions within the body, recent work has caused new biomaterials with promise to solve orthopedic problems [36].

Table 2. Some interesting reports on applied usage of nanomaterials for orthopedics

Authors	Details
Ni et al. [37]	In this study, a bioactive bone cement, which consists of of strontium-consist ofing hydroxyapatite (Sr-HA) powder and bisphenol-alpha-glycidyl dimethacrylate (Bis-GMA)-based resin, was assessed in an animal model [37]. According to this study, osseointegration of Sr-HA cement with cancellous bone was observed and apatite layer with high content of calcium and phosphorus was found between cancellous bone and Sr-HA cement [37]. Ni et al. indicated that these two interfaces were diffused interfaces, and cancellous bone or cortical bone was developed into Sr-HA cement 6 months after the implantation [37].
Chris Arts et al. [38]	Cris Arts et al. reported on the use of a bioresorbable nano-crystalline hydroxyapatite paste in acetabular bone impaction grafting [38]. In this work, destructive lever-out tests and in vivo animal tests were performed with various combinations of materials [38]. Chris Arts et al. summarized that TCP-HA granules with a nano-crystalline hydroxyapatite paste could be a valuable addition when TCP-HA ceramic granules are being used for acetabular bone impaction grafting procedures [38].
Sun et al. [39]	Sun et al. performed a study to investigate the effect of nano-hydroxyapatite/collagen (nHA/collagen) composite as a graft extender and enhancer when combined with recombinant human bone morphogenetic protein-2 (rhBMP-2) on lumbar intertransverse fusion in rabbits [39]. Sun et al. concluded that the nHA/collagen, especially combined with rhBMP-2, was a useful bone substitute, for it has rapid biodegradation, fine bone-bending ability, and high osteoconductivity on posterolateral spinal fusion in rabbits [39].

NANODENTISTRY [40]

Similar to nanosurgery, applied usage of nanobiotechnology to nanodentistry can be early described as the development of many tools and devices. Nanotechnology also has a significant involvement in dentistry materials. Some interesting reports on nanodentistry are quoted in Table 3.

Table 3. Some interesting reports on nanodentistry

Authors	Details
Schirrmeister et al. [41]	The purpose of this work was to assess the clinical characteristic of a new restorative material (Ceram.X) in combination with a new primer-adhesive (K-0127) [41]. A measurable result from actual usage could be observed in this work [41].
He and Swain [42]	In this work, a nano-based indentation system (Ultra Micro-Indentation System, UMIS-2000, CSIRO, Australia) was used to determine the indentation stress-strain response of two kinds of dental ceramics, one kind of dental alloy and healthy enamel [42]. He and Swain said that strong and tough materials with primarily elastic response, such as toughened ceramics, were required to enable dental crown/bridges to have long-term reliability [42].

Authors	Details
Lu et al. [43]	This study compared the mechanical properties, generalized wear resistance and polymerization shrinkage of a resin composite filled with spherical inorganic filler to other commercial resin composites including nano-composite [43]. Lu et al. found that all the materials had a similar shrinkage pattern in that about 99% of shrinkage occurred in less than 24 hours [43].
Bishara et al. [44]	Bishara et al. performed this work to compare the shear bond strength of a nano-hybrid restorative material, Grandio, to that of a traditional adhesive material when bonding orthodontic brackets. Bishara et al. concluded that the newly introduced nano-filled composite materials could potentially be used to bond orthodontic brackets to teeth if its consistency could be more flowable to readily adhere to the bracket base [44].
Croll and Breg [45]	In this work, a nano-filled resin-modified glass-ionomer was introduced for filling dental caries [45].
Fu et al. [46]	Fu et al. studied effects of dental bleaching on micro- and nano-morphological alterations of the enamel surface [46]. Fu et al. concluded that the thickness of the enamel smear layer was significantly reduced due to the bleaching process [46].
Hairul Nizam et al. [47]	Hairul Nizam et al. performed a nanoindentation study of human premolars subjected to bleaching agent [47]. Hairul Nizam et al. said that the exact mechanism by which hydrogen peroxide impacts the dentin and enamel had yet to be completely elucidated; however, it was observed to have an undermining effect on the nanomechanical properties of teeth [47].
Lee et al. [48]	The objective was to assess the color changes after curing, polishing, and thermocycling of a nano-filled resin composite [48]. Lee et al. concluded that changes of color and translucency after curing, polishing, and thermocycling varied by the shade group [48].
Van Meerbeek et al. [49]	Assessment by nano-indentation of the hardness and elasticity of the resin-dentin bonding area was performed in this work [49]. Van Meerbeek et al. concluded that such an elastic bonding area might have a strain capacity sufficient to relieve stresses between the shrinking composite restoration and the rigid dentin substrate, thereby improving the conservation of the dentin bond and, as a consequence, the marginal integrity and retention of the restoration [49].
Suzuki [50]	This study assessed the relative wear resistance of many types of denture teeth using an in vitro wear testing device. Suzuki concluded that The nano-composite tooth was harder and more wear-resistant than the acrylic teeth but not significantly different from most of the cross-linked and micro-filled composite teeth tested [50].

Gene Therapy and Nanobiotechnology

Gene therapy is the advanced therapeutic concept at present. By definition, gene therapy implies local or systemic administration of a nucleic acid construct that can prevent, treat and even cure diseases by changing the expression of genes that are responsible for the pathological condition. Gene therapy involves gene manipulation and transfer. A number of polymer-based synthetic systems, or 'vectors', have been developed to entice cells to use exogenous DNA [51]. This therapy is the hope for treatment of presently untreatable diseases. Gene-delivery systems usually include a carrier system which both protects the gene expression plasmid and allows its extracellular and intracellular transportation [52]. Viruses are used in most of the clinical experiments today; however, they do have important drawbacks. Therefore, non-viral vectors based on lipids, hydro-soluble polycations, other non-condensing polymers and nano- or microparticles/capsules have been proposed [52]. Nanobiotechnology can be applied for the complex gene therapy process. At present, both biodegradable and non-biodegradable inorganic particles can be completely fabricated on the nano-scale with the attributes of binding DNA, internalizing across the plasma membrane and finally releasing it in the cytoplasm for final expression of a protein [53]. In addition to the classical system (intravenous injection), development of polymer-based nanoparticle technologies for oral gene therapy is in continous development [54]. Some interesting reports on gene therapy and nanobiotechnology are quoted in Table 4.

Table 4. Some interesting reports on gene therapy and nanobiotechnology

Authors	Details
Pan et al. [55]	In this work, Pan et al. revealed that polyamidoamine dendrimer-modified MNPs might be a good gene delivery system and have potential applied usages in cancer therapy and molecular imaging diagnosis [55].
Hatakeyama et al. [56]	In this work, Hatakeyama et al. reported on development of a new systemic gene delivery system for cancer therapy with a tumor-specific cleavable PEG-lipid [56]. The development of a PEG-peptide-lipid ternary conjugate was completely developed in this study [56]
Kogure [57]	In this work, a multifunctional envelope-type nano device (MEND) was developed for use as an efficient nonviral system for the delivery of plasmid DNA (pDNA), oligodeoxynucleotide (ODN), and siRNA using octaarginine (R8) as an internalizing ligand based on the programmed packaging [57]. Kogure said that MEND was a useful new delivery system for pDNA and functional nucleic acids [57].
Hosseinkhani and Tabata [58]	Hosseinkhani and Tabata reported on self assembly of DNA nanoparticles with polycations for the delivery of genetic materials into cells [58]. This study clearly demonstrated that self-integrated DNA complexes had potential as a gene delivery vehicle and were stable to transfer genetic materials to cells [58].
Hatakeyama et al. [59]	In this work, Hatakeyama et al. reported on development of a new systemic gene delivery system for cancer therapy with a tumor-specific cleavable PEG-lipid [59].
Yamada et al. [60]	Yamada et al. noted that HBsAg L particles were able to deliver therapeutic payloads with high specificity to human hepatocytes [60]. Yamada et al. indicated that the L particle was suitable for a cell- and tissue-specific gene/medicine transfer vector [60].

Authors	Details
Kleemann et al. [61]	Kleeman et al. described a new modification of PEI, in which an oligopeptide related to the protein transduction domain of HIV-1 TAT was covalently coupled to 25 kDa PEI (PEI) through a heterobifunctional polyethylenglycol (PEG) spacer resulting in a TAT-PEG-PEI conjugate [61]. Kleeman et al. proposed that their new modification represented a new approach to non-viral gene carriers for lung therapy, comprising protection for plasmid DNA, low toxicity and significantly amplified transfection efficiency under in vivo conditions [61].

Stem Cell Therapy and Nanobiotechnology

Stem cell therapy is another advanced therapeutic concept. It is applied for treatment of many diseases especially for congenital defects and malignancies. The best examples of nanobiotechnology applied usage in stem cell therapy can be seen in nanohematology [62 - 65]. More details can be seen in the chapter on nanohematology in this book.

Tissue Engineering and Nanobiotechnology

Tissue engineering is the advanced biotechnology that is useful for regenerative medicine [66]. Basically, cross-linking and cross-bridging are highly versatile methods of creating composite protein structures with desired mechanical properties such as deformation endurance, elasticity, extensibility, and stability under intensive and repetitive sheering forces [67]. The nanomaterials can be used as structural cross-linking and cross-bridging polymers for the newly-generated tissues. With the ability to form nano-fibrous structures, a drive to mimic the extracellular matrix and form scaffolds that are an artificial extracellular matrix suitable for tissue formation has begun [68 - 69]. Smith and Ma proposed that the nano-fibrous scaffolds attempted to mimic collagen, a natural extracellular matrix component, and could potentially bring a better environment for tissue formation in tissue engineering systems [68]. Murugan and Ramakrishna noted that scaffolds played a pivotal role in supporting the cells to accommodate and guide their growth into a specific tissue; therefore, designing scaffolds that are favorable for cellular growth was important [70]. Murugan and Ramakrishna also said that electrospinning was a straightforward, cost-effective technique that could be applied to the fabrication of nano-featured scaffolds suitable for tissue engineering, and it offered usefulness over conventional scaffold methodologies [70]. Some interesting reports on tissue engineering and nanobiotechnology are quoted in Table 5.

Table 5. Some interesting reports on tissue engineering and nanobiotechnology

Authors	Details
Chen et al. [71]	Chen et al. performed a meta-analysis of applied usage of electrostatic spinning technology in nano-structured polymer scaffold [71]. Chen et al. concluded that the nano-structured polymer scaffold could support the cell adhesion, proliferation, site, and differentiation, and this kind of scaffold had a considerable value in the tissue engineering field [71].
Srouji et al. [72]	This study examined the applied usage of a new 3-D construct, produced by multilayered organization of electrospun nanofiber membranes, for human bone marrow-derived mesenchymal stem cells (hMSCs) support [72]. Srouji et al. concluded that the combination of electrospinning technology with multilayer technique produced the new 3-D nanofiber multilayered construct, able to consist of efficient cell mass necessary for a full in vivo grafting [72]. Srouji et al. also mentioned that the success of this protocol with undifferentiated cells implied the possibility of its applied usage as a platform for development of constructs with cells directed into various tissue types [72].
Ren et al. [73]	In this study, Poly (D: ,L: -lactide)/nano-hydroxyapatite (PDLLA/NHA) composites were treated by in-situ polymerization, and highly porous scaffolds were fabricated using a new method, supercritical CO(2)/salt-leaching method (SC CO(2)/SL) [73].
Haung et al. [74]	Haung et al. reported their development of nano-sized hydroxyapatite-reinforced composites for tissue engineering scaffolds [74].
Ren et al. [75]	In this work, PLLA, PLA-PEG and PLGA porous scaffolds with pore sizes ranging from 100 to 250 microns and porosity over 85% were fabricated by a solution-casting/salt-leaching method [75]. Ren et al. found that these composite scaffolds can be a potentially effective tool for treating mandibular and other bone defects [75].
Alsberg et al. [76]	In this work, Alsberg et al. described a method for fabricating fibrin gels with defined architecture on the nanometer scale in which magnetic forces were used to position thrombin-coated magnetic micro-beads in a defined 2-dimensional array and thereby guide the self-assembly of fibrin fibrils through catalytic cleavage of soluble fibrinogen substrate [76].

NANOBIOSENSOR IMPLANT

Nanobiosensor implant is an applied usage of a nanobiodevice. At present, it is accepted that it is possible to implant a very small, nanoscale, biodevice into the human body. The nanobiosensor implant can be a useful for treatment aiming at physiological response to the rhythmic change in the human body. The best example is the case of diabetes mellitus, which is an endocrine disorder [77 – 78]. Basically, diabetes mellitus is a disorder caused by the insufficient production of the pancreatic hormone, insulin. The treatment must match with the daily physiological rhythmic change of insulin in the human being.

REFERENCES

[1] Murthy SK. Nanoparticles in novel medicine: state of the art and future challenges. *Int. J. Nanomedicine.* 2007;2(2):129-41.

[2] Pauwels EK, Erba P Towards the use of nanoparticles in cancer therapy and imaging. *Medicine News Perspect.* 2007 May;20(4):213-20.

[3] Moghimi SM, Hunter AC, Murray JC. Nanomedicine: present status and future prospects. *FASEB J.* 2005 Mar;19(3):311-30.

[4] Emerich DF. Nanomedicine--prospective therapeutic and diagnostic applied usages. *Expert Opin. Biol Ther.* 2005 Jan;5(1):1-5.

[5] Nakayama M, Okano T. Medicine delivery systems using nano-sized medicine carriers. *Gan To Kagaku Ryoho.* 2005 Jul;32(7):935-40.

[6] Erickson RP. From "magic bullet" to "specially engineered shotgun loads": the new genetics and the need for individualized pharmacotherapy. *Bioessays.* 1998 Aug;20(8):683-5.

[7] Osieka R. Oncology. *Arzneimittelforschung.* 1997 Oct;47(10):1161-5.

[8] Galanski M, Keppler BK. Searching for the magic bullet: anticancer platinum medicines which can be accumulated or activated in the tumor tissue. *Anticancer Agents Med. Chem.* 2007 Jan;7(1):55-73.

[9] Welch DR. Biologic considerations for medicine targeting in cancer patients. *Cancer Treat. Rev.* 1987 Dec;14(3-4):351-8.

[10] Kreuter MW. Human property and cancer. Forobtain the magic bullet! *Cancer.* 1993 Aug 1;72(3 Suppl):996-1001.

[11] Couvreur P, Vauthier C. Nanotechnology: intelligent design to treat complex disease. *Pharm. Res.* 2006 Jul;23(7):1417-50.

[12] Chen HC. New era of reconstructive microsurgery. *Changgeng Yi Xue Za Zhi.* 1997 Sep;20(3):153-62.

[13] Chicarilli ZN. Pediatric microsurgery: revascularization and replantation. *J. Pediatr. Surg.* 1986 Aug;21(8):706-10.

[14] Arnold DJ, Wax MK; Microvascular Committee of the American Academy of Otolaryngology--Head and Neck Surgery. Pediatric microvascular reconstruction: a report from the Microvascular Committee. *Otolaryngol. Head Neck Surg.* 2007 May;136(5):848-51.

[15] Strasburger VC. First do no harm: why have parents and pediatricians missed the boat on children and media? *J. Pediatr.* 2007 Oct;151(4):334-6.

[16] Smoyak SA. Primum non nocere (first, do no harm). *J. Psychosoc. Nurs Ment. Health Serv.* 2007 Sep;45(9):8-9.

[17] Evans J. First, do no harm. *Pract. Midwife.* 2007 Sep;10(8):22-3.

[18] Monreal M, Trujillo-Santos J. Emerging Concepts in Clinical Practice Guidelines. Z *Gastroenterol.* 2007 Oct;45(10):1075-1081.

[19] Dutta RC. Medicine carriers in pharmaceutical design: promises and progress. *Curr. Pharm. Des.* 2007;13(7):761-9.

[20] Dobson J. Magnetic micro- and nano-particle-based targeting for medicine and gene delivery. *Nanomed.* 2006 Jun;1(1):31-7.

[21] Freitas RA Jr. Nanotechnology, nanomedicine and nanosurgery. *Int. J. Surg.* 2005;3(4):243-6. Epub 2005 Nov 28.
[22] Ebbesen M, Jensen TG. Nanomedicine: ways, potentials, and ethical implications. *J. Biomed. Biotechnol.* 2006;2006(5):51516.
[23] Fankhauser F, Niederer PF, Kwasniewska S, van der Zypen E. Supernormal vision, high-resolution retinal imaging, multiphoton imaging and nanosurgery of the cornea--a review. *Technol. Health Care.* 2004;12(6):443-53.
[24] Tirlapur UK, Konig K. Femtosecond near-infrared laser pulses as a versatile non-invasive tool for intra-tissue nanoprocessing in plants without comuseful viability. *Plant J.* 2002 Aug;31(3):365-74.
[25] Firtel M, Henderson G, Sokolov I. Nanosurgery: observation of peptidoglycan strands in Lactobacillus helveticus cell walls. *Ultramicroscopy.* 2004 Nov;101(2-4):105-9.
[26] Sacconi L, Tolic-Norrelykke IM, Antolini R, Pavone FS. Combined intracellular three-dimensional imaging and selective nanosurgery by a nonlinear microscope. *J. Biomed. Opt.* 2005 Jan-Feb;10(1):14002.
[27] Kohli V, Elezzabi AY, Acker JP. Cell nanosurgery using ultrashort (femtosecond) laser pulses: applied usages to membrane surgery and cell isolation. *Lasers Surg. Med.* 2005 Sep;37(3):227-30.
[28] Colombelli J, Reynaud EG, Rietdorf J, Pepperkok R, Stelzer EH. In vivo selective cytoskeleton dynamics quantification in interphase cells induced by pulsed ultraviolet laser nanosurgery. *Traffic.* 2005 Dec;6(12):1093-102.
[29] Wang BG, Halbhuber KJ. Corneal multiphoton microscopy and intratissue optical nanosurgery by nanojoule femtosecond near-infrared pulsed lasers. *Ann. Anat.* 2006 Sep;188(5):395-409.
[30] Jeffries GD, Edgar JS, Zhao Y, Shelby JP, Fong C, Chiu DT. Using polarization-shaped optical vortex traps for single-cell nanosurgery. *Nano Lett.* 2007 Feb;7(2):415-20.
[31] Wakida NM, Lee CS, Botvinick ET, Shi LZ, Dvornikov A, Berns MW. Laser nanosurgery of single microtubules reveals site-dependent depolymerization rates. *J. Biomed. Opt.* 2007 Mar-Apr;12(2):024022.
[32] Colombelli J, Reynaud EG, Stelzer EH. Investigating relaxation processes in cells and producing organisms: from cell ablation to cytoskeleton nanosurgery. *Methods Cell Biol.* 2007;82:267-91.
[33] Leary SP, Liu CY, Apuzzo ML. Toward the emergence of nanoneurosurgery: part II--nanomedicine: diagnostics and imaging at the nanoscale level. *Neurosurgery.* 2006 May;58(5):805-23.
[34] Leary SP, Liu CY, Apuzzo ML. Toward the emergence of nanoneurosurgery: part III--nanomedicine: targeted nanotherapy, nanosurgery, and progress toward the realization of nanoneurosurgery. *Neurosurgery.* 2006 Jun;58(6):1009-26
[35] Leary SP, Liu CY, Apuzzo ML. Toward the emergence of nanoneurosurgery: part I--progress in nanoscience, nanotechnology, and the comprehension of events in the mesoscale realm. *Neurosurgery.* 2005 Oct;57(4):606-34
[36] Balasundaram G, Webster TJ. An overview of nano-polymers for orthopedic applied usages. *Macromol. Biosci.* 2007 May 10;7(5):635-42.

[37] Ni GX, Choy YS, Lu WW, Ngan AH, Chiu KY, Li ZY, Tang B, Luk KD. Nanomechanics of bone and bioactive bone cement interfaces in a load-bearing model. *Biomaterials.* 2006 Mar;27(9):1963-70.

[38] Chris Arts JJ, Verdonschot N, Schreurs BW, Buma P. The use of a bioresorbable nano-crystalline hydroxyapatite paste in acetabular bone impaction grafting. *Biomaterials.* 2006 Mar;27(7):1110-8.

[39] Sun TS, Guan K, Shi SS, Zhu B, Zheng YJ, Cui FZ, Zhang W, Liao SS. Effect of nano-hydroxyapatite/collagen composite and bone morphogenetic protein-2 on lumbar intertransverse fusion in rabbits. *Chin. J. Traumatol.* 2004 Feb;7(1):18-24.

[40] Freitas RA Jr. Nanodentistry. J Am Dent Assoc. 2000 Nov;131(11):1559-65.

[41] Schirrmeister JF, Huber K, Hellwig E, Hahn P. Two-year evaluation of a new nano-ceramic restorative material. *Clin. Oral Investig.* 2006 Sep;10(3):181-6.

[42] He LH, Swain MV. Nanoindentation derived stress-strain properties of dental materials. *Dent Mater.* 2007 Jul;23(7):814-21.

[43] Lu H, Lee YK, Oguri M, Powers JM. Properties of a dental resin composite with a spherical inorganic filler. *Oper. Dent.* 2006 Nov-Dec;31(6):734-40.

[44] Bishara SE, Ajlouni R, Soliman MM, Oonsombat C, Laffoon JF, Warren J. Evaluation of a new nano-filled restorative material for bonding orthodontic brackets. *World J. Orthod.* 2007 Spring;8(1):8-12.

[45] Croll TP, Berg JH.c Resin-modified glass-ionomer restoration of primary molars with proximating Class II caries lesions. *Compend Contin. Educ. Dent.* 2007 Jul;28(7):372-6.

[46] Fu B, Hoth-Hannig W, Hannig M. Effects of dental bleaching on micro- and nano-morphological alterations of the enamel surface. *Am. J. Dent.* 2007 Feb;20(1):35-40.

[47] Hairul Nizam BR, Lim CT, Chng HK, Yap AU. Nanoindentation study of human premolars subjected to bleaching agent. *J. Biomech.* 2005 Nov;38(11):2204-11.

[48] Changes of optical properties of dental nano-filled resin composites after curing and thermocycling. *J. Biomed. Mater Res. B Appl. Biomater.* 2004 Oct 15;71(1):16-21.

[49] Van Meerbeek B, Willems G, Celis JP, Roos JR, Braem M, Lambrechts P, Vanherle G. Assessment by nano-indentation of the hardness and elasticity of the resin-dentin bonding area. *J. Dent. Res.* 1993 Oct;72(10):1434-42.

[50] Suzuki S. In vitro wear of nano-composite denture teeth. *J. Prosthodont.* 2004 Dec;13(4):238-43.

[51] Putnam D. Polymers for gene delivery across length scales. *Nat Mater.* 2006 Jun;5(6):439-51.

[52] Piskin E, Dincer S, Turk M. Gene delivery: intelligent but just at the beginning. *J. Biomater. Sci. Polym. Ed.* 2004;15(9):1181-202.

[53] Chowdhury EH, Akaike T. Bio-functional inorganic materials: an attractive branch of gene-based nano-medicine delivery for 21st century. *Curr Gene Ther.* 2005 Dec;5(6):669-76.

[54] Bhavsar MD, Amiji MM. Polymeric nano- and microparticle technologies for oral gene delivery. *Expert. Opin. Medicine. Deliv.* 2007 May;4(3):197-213.

[55] Pan B, Cui D, Sheng Y, Ozkan C, Gao F, He R, Li Q, Xu P, Huang T. Dendrimer-modified magnetic nanoparticles enhance efficiency of gene delivery system. *Cancer Res.* 2007 Sep 1;67(17):8156-63.

[56] Hatakeyama H, Akita H, Kogure K, Harashima H. Development of a new systemic gene delivery system for cancer therapy with a tumor-specific cleavable PEG-lipid. *Yakugaku Zasshi.* 2007 Oct;127(10):1549-56.

[57] Kogure K.Development of a new artificial gene delivery system multifunctional envelope-type nano device for gene therapy. *Yakugaku Zasshi.* 2007 Oct;127(10):1685-91.

[58] Hosseinkhani H, Tabata Y. Self assembly of DNA nanoparticles with polycations for the delivery of genetic materials into cells. *J. Nanosci. Nanotechnol.* 2006 Aug;6(8):2320-8.

[59] Hatakeyama H, Akita H, Kogure K, Oishi M, Nagasaki Y, Kihira Y, Ueno M, Kobayashi H, Kikuchi H, Harashima H. Development of a new systemic gene delivery system for cancer therapy with a tumor-specific cleavable PEG-lipid. *Gene* Ther. 2007 Jan;14(1):68-77.

[60] Yamada T, Ueda M, Seno M, Kondo A, Tanizawa K, Kuroda S. New tissue and cell type-specific gene/medicine delivery system using surface engineered hepatitis B virus nano-particles. *Curr. Medicine. Targets Infect .Disord.* 2004 Jun;4(2):163-7.

[61] Kleemann E, Neu M, Jekel N, Fink L, Schmehl T, Gessler T, Seeger W, Kissel T. Nano-carriers for DNA delivery to the lung based upon a TAT-derived peptide covalently coupled to PEG-PEI. *J. Control. Release.* 2005 Dec 5;109(1-3):299-316.

[62] Chen H, Titushkin I, Stroscio M, Cho M. Altered membrane dynamics of quantum dot-conjugated integrins during osteogenic differentiation of human bone marrow derived progenitor cells. *Biophys J.* 2007 Feb 15;92(4):1399-408.

[63] Garon EB, Marcu L, Luong Q, Tcherniantchouk O, Crooks GM, Koeffler HP. Quantum dot tagging and tracking of human leukemic, bone marrow and cord blood cells. *Leuk. Res.* 2007 May;31(5):643-51.

[64] Horak D, Babic M, Jendelova P, Herynek V, Trchova M, Pientka Z, Pollert E, Hajek M, Sykova E. D-mannose-modified iron oxide nanoparticles for stem cell tagging. *Bioconjug. Chem.* 2007 May-Jun;18(3):635-44.

[65] Zheng J, Ghazani AA, Song Q, Mardyani S, Chan WC, Wang C. Cellular imaging and surface marker tagging of hematopoietic cells using quantum dot bioconjugates. *Lab. Hematol.* 2006;12(2):94-8.

[66] West JL. "Nano-technologies for biomedical applied usages". Conf Proc IEEE Eng Med *Biol. Soc.* 2006;1:nil19-20.

[67] Levy I, Shoseyov O. Cross bridging proteins in nature and their utilization in bio- and nanotechnology. *Curr. Protein Pept. Sci.* 2004 Feb;5(1):33-49.

[68] Smith LA, Ma PX. Nano-fibrous scaffolds for tissue engineering. *Colloids Surf. B Biointerfaces.* 2004 Dec 10;39(3):125-31.

[69] Kikuchi A, Okano T. Nanostructured designs of biomedical materials: applied usages of cell sheet engineering to functional regenerative tissues and organs. *J. Control Release.* 2005 Jan 3;101(1-3):69-84.

[70] Murugan R, Ramakrishna S. Nano-featured scaffolds for tissue engineering: a review of spinning methodologies. *Tissue Eng.* 2006 Mar;12(3):435-47.

[71] Chen D, Li M, Fang Q. Applied usage of electrostatic spinning technology in nano-structured polymer scaffold. *Zhongguo Xiu Fu Chong Jian Wai Ke Za Zhi.* 2007 Apr;21(4):411-5.

[72] Srouji S, Kizhner T, Suss-Tobi E, Livne E, Zussman E. 3-D Nanofibrous electrospun multilayered construct is an alternative ECM mimicking scaffold. *J. Mater Sci. Mater Med.* 2007 Aug 15;

[73] Ren J, Zhao P, Ren T, Gu S, Pan K. Poly (D: ,L: -lactide)/nano-hydroxyapatite composite scaffolds for bone tissue engineering and biocompatibility evaluation. *J Mater Sci Mater Med.* 2007 Aug 15;

[74] Huang J, Lin YW, Fu XW, Best SM, Brooks RA, Rushton N, Bonfield W. Development of nano-sized hydroxyapatite reinforced composites for tissue engineering scaffolds. *J. Mater Sci. Mater Med.* 2007 Sep 20;

[75] Ren J, Ren T, Zhao P, Huang Y, Pan K. Repair of mandibular defects using MSCs-seeded biodegradable polyester porous scaffolds. *J. Biomater. Sci. Polym. Ed.* 2007;18(5):505-17.

[76] Alsberg E, Feinstein E, Joy MP, Prentiss M, Ingber DE. Magnetically-guided self-assembly of fibrin matrices with ordered nano-scale structure for tissue engineering. *Tissue Eng.* 2006 Nov;12(11):3247-56.

[77] Staples M, Daniel K, Cima MJ, Langer R. Applied usage of Micro- and Nano-Electromechanical Devices to Medicine Delivery. *Pharm. Res.* 2006 May 5;

Chapter 7

NANODIAGNOSIS

WHAT IS NANODIAGNOSIS?

The genesis of nanotechnology can be traced to the promise of revolutionary advances across medicine, communications, genomics and robotics [1]. Nanodiagnosis is a diagnostic way making use of nanoparticles. Nanomaterials and nanotechnology combined with novel devices have the potential to address this emerging challenge [2]. Using a multiple appearance of nanomaterials for multiplex diagnostics and imaging applications will offer sensitive, rapid, and cost-effective solutions for the modern clinical laboratory [2]. Medvedeva et al. proposed that implementation of optico-biosensoric, atomic-force, nanowire and nanoporous approaches into genomics and proteomics could significantly enhance the sensitivity and accuracy of diagnostics and would shorten the time of diagnostic procedures that will undoubtedly improve the efficiency of medical treatment [3].

Basically, different nanomaterials can be applied for medical diagnostic purposes. Among the different nanomaterials used to construct nanoparticles are organic polymers, co-polymers, and metals [4]. Some of these materials can self assemble, and depending on the conditions under which the self-assembly process occurs, a vast array of shapes can be formed [4]. Frequently, the nanoparticle morphology is spherical or tubular, mimicking the shape but, thus far, not the functions of subcellular organelles [4]. Rotomskis et al. said that the properties of nanomaterials (especially quantum and surface area effects) could improve the sensitivity of biological detection and imaging by at least 10- to 100-fold [5]. In this chapter, important aspects of nanodiagnosis will be summarized and discussed.

NAKED EYE DIAGNOSTIC SYSTEM BY NANOPARTICLE

The most simple technique in diagnosis can bring wide generalization among users. The naked eye system is an example of a simple test that can be used as a bedside test. Indeed, the naked eye system is not new in laboratory medicine. There are many classical diagnostic tests that can be classified as members of the naked eye system. The protein precipitation systems to qualitatively test the protein in body fluids are the best example. For detection of protein in body fluid, the nanoparticles can also be applied to setting up a naked eye system. Basically, the nanoparticle has many interesting properties. An important property is the changing of color due to the aggregation of the particles [6]. This can be applied to diagnostic tests in medicine. In this situation, the binding between the positive charge within the particle and negative charge within the microalbumin molecule in a positive sample can be expected, while this reaction will not occur in a negative sample. Wiwanitkit et al. recently reported the ful applied usage of gold nanoparticle solution as a naked eye system for detection of microalbuminuria [7]. Normal urine triggers the precipitation of gold nanoparticle solution resulting in a grey coloured mixture, while urine samples consisting of protein do not. For further generalization, the author hereby presented additional result on the test on the feasibility of this system for detection of proteinuria. In this experiment, following the classical Turkevich citrate reduction method [8], nine nanometer-sized gold nanoparticle solutions were synthesized. An AuNP solution was treated at a concentration of 44 ppm. The gold nanoparticle solution can be stored in the dark under 4°C for over a month. $HAuCl_4 \cdot 3H_2O$ was purchased from Arcos Organic and trisodium citrate dihydrate was purchased from Merck Ltd. The test protocol was divided into two parts: a test for the feasibility in a real urine specimen and an additional study for detection limit and specificity of the test. Considering the test for the feasibility in a real urine specimen, 250μL samples of urine were pipetted into individual cuvettes The study included urine samples of, for example, glucosuria, ketonuria, proteinuria and a normal urine sample which were characterised via standard routine automated urinalysis. The characteristics of tested samples are summarised in Table 1. Then 250μL of the gold nanoparticle solution were individually mixed into each cuvette. All solution mixtures were then continuously shaken for 10 minutes. The existence/absence of gold nanoparticle aggregation was warranted by the colour change of the solution mixture as mentioned above. The AuNP solution, urine samples, and the solution mixtures were tested for UV-visible absorption using a Varian® UV-visible spectrophotometer, CARY 50 Probe. Concerning additional study for detection limit and specificity of the test, synthetic urine was treated according to the previous report by Tanaka and Hayashi [9]. The detection limit was detected by UV-visible titration between gold nanoparticles and BSA consisting of synthetic urine. First, UV-visible absorption spectrum of 500 μL of gold nanoparticles was recorded. Synthetic urine with various concentrations of bovine serum albumin (BSA) were then added and the UV-Visible absorption spectra of each addition were recorded consecutively. The specificity of this test was assessed using BSA consisting of synthetic urine at the concentration of the detection limit. The 191 samples of 50 μL BSA-urine were pipetted into multiple well plates, then 100 μL of gold nanoparticles were added into each well. The colour of the mixtures was observed.

Table 1. Characteristics of tested urine samples

Urine sample	Proteinuria	Hematuria	Haemoglobinuria	Ketonuria
b	-	-	-	-
c	+	+	+	+
d	+	+	-	-
e	+	-	+	+

The gold nanoparticles were mixed with urine samples for 10 minutes and the colour change was then recorded. All solution mixtures containing protein exhibited no colour change. On the other hand, the normal urine without protein turned the solution mixture from pink to grey. The gold nanoparticle method was proven to give consistent positive or negative results with the validated automated-urinalysis analyzer in all samples. This observation indicates that protein effectively prevents gold nanoparticles from aggregating in urine. The normal urine sample exhibited aggregation of gold nanoparticles, indicating that the environment in the sample is not suitable for sustaining the particle distribution. From 25 µg/mL of BSA in urine and higher, the colour of gold nanoparticles did not change to grey, indicating that the aggregation of gold nanoparticles can be inhibited for both gold nanoparticles: Urine = 100:25 µL (top) and 100:50 µL (bottom). Hence, the detection limit of BSA in urine by this method is 25 µg/mL. Absorption Maxima of gold nanoparticles-BSA/Urine mixture indicated that the gold nanoparticles' colour was completely shifted from 520 nm to 680 nm approximately with BSA-free urine. Hence, the colour of gold nanoparticles was switched from pink to grey. At BSA concentration 2-20 µg/mL, the broad absorption yielded the grey colour of gold nanoparticles. With BSA concentration 25 µg/mL and above, the colour of the solution mixture remained pink, suggesting that the aggregation of gold nanoparticles was inhibited. This gold nanoparticle assay for urine protein can be performed in a single tube with a total work-up time of 10 minutes. The procedure is easy and requires simple laboratory equipment. The method is considerably less expensive than other methods. Additionally, the detection limit is 25 µg/mL. For our knowledge, this result shows that our method is so far the most effective detection in low concentration proteinuria. In conclusion, this work demonstrated that protein molecules play a key role in preventing AuNPs from precipitating in urine. The gold nanoparticle solution is a rapid and sensitive tool for the visual detection of protein in urine through a clear colour change, without any modification needed. Thus, we introduce unmodified gold nanoparticles as a new tool for protein screening in urine.

Figure 1. Sample with proteinuria.

In addition to detection of protein in body fluid, the naked eye system can also be applied to detection of some small-scale substances in body fluid such as hormones. To help the reader better understand, the author hereby presents some preliminary reports on an experimental study to test the feasibility of gold nanoparticle solution for detection of urine human choriogonadotropin (hCG), which is a basic commonly used diagnostic test for screening for pregnancy in females with missed menstruation. At present, a urine pregnancy test is routinely performed with a urine pregnancy strip test. The test is performed based on an immunochromatography method. Here, the authors used the advanced nanomedicine way for modification of the urine pregnancy test. Of interest, it can be seen that gold nanoparticles can be a useful alternative tool for a urine pregnancy test. This study was a laboratory experiment. The remaining samples from routine analysis for urine pregnancy test were used. The author performed an experiment on both pregnancy positive and negative urine samples. In this study, the positive urine sample consisted of hCG level equal to 50 mIU/mL. First, a mixture of equal summations (500 microliter) of gold nanoparticle solution and urine sample was treated. Then it was further tested for the urine pregnancy test by urine pregnancy test strip (WH Accu Test by W.H.P.M. Inc EI Monte, CA; the sensitivity of the test is 100%). The defining characteristics of this new gold nanoparticle system was examined for its accuracy, specificity, sensitivity, imprecision, stability, interfering substances, expected value, calibration, and limitations according to the guideline protocol of the Food Medicine Administration of the United States (FDA – 2001). For analysis of accuracy, specificity and sensitivity, 2100 samples were analyzed per parameters. For imprecision analysis, thirty samples were analyzed. For stability, repeated analysis was performed for one week. For interfering substances, the effect of glucose and ketone was tested. To study the expected value, calibration, and limitations of the system, the detection limit was detected by UV-visible titration between gold nanoparticle solution and hCG added urine. First, UV-visible absorption spectrum of 500 μL of gold nanoparticle solution was recorded. Synthetic urine with various concentrations of hCG were then added and the UV-visible absorption spectra of each addition were recorded consecutively. UV-visible absorption using a Varian® UV-visible spectrophotometer, CARY 50 probe was used. The positive and negative urine samples were tested by urine pregnancy strips (two lines for positive and one line for negative). The mixtures of gold nanoparticle solution and positive and negative urine samples were treated. The pregnancy positive mixture became pink while the pregnancy negative mixture turned gray. The urine pregnancy test strip for positive mixture had two lines while the negative mixture had one line. Considering the examination for defining characteristic, accuracy, specificity and sensitivity was equal to 100%, 100%, 100% and 100%, respectively. Considering imprecision analysis, precision was equal to 100 %. For stability, the test was stable for one week test period. For interfering substances, there was no effect of glucose or ketone on the system. According to the study on expected value, calibration, and limitations, from 50 mIU/mL of hCG in urine and higher, the color of gold nanoparticle did not change to grey as shown in Figure 1. Hence, the detection limit of hCG in urine by this method is 50 mIU/mL. In addition, at the studied level ranging from 50 to 500 mIU/mL, the linearity of the tested can be observed. In this situation, the binding between the positive charge within the particle and negative charge within the hCG in the positive urine sample can be expected, while this reaction will not occur in the negative urine sample. This is believed to be the explanation for the difference in color. Indeed, the theory on metallic nanoparticles that these particles can offer control over the size and protection against aggregation with relation to color change is warranted [10]. A similar colorimetric system for blood determinants was proposed in February 2007 [11]. This applied usage can help diagnosis of pregnancy and can be the alternative method for this urine pregnancy test. In addition, the cost of the gold nanoparticle (about 0.2 US dollar in Thailand) is cheaper than the urine strip test (about 1 US dollar in Thailand).

Recently, Bauer et al. also reported another interesting naked eye diagnostic system by nanoparticle [12]. Based on the understanding of the absorption property of metal nanoparticles, Bauer et al. aimed at the direct detection of sub-monomolecular layers of DNA with the naked eye [12]. Bauer et al. focused on the nanolayer-coated metallized-PET-chip setup and on the synthesis of DNA-nanoparticle conjugates suitable for 'resonance amplified absorption'-point of care-tests and the applied usage of those particles in the direct visualization of DNA-DNA binding events [12]. A result with practical implications could be derived in this work [12].

Immunological Diagnostic System by Nanoparticle

Immunodiagnosis by nanoparticle is the new phase of immunological test at present. Efforts to improve the characteristic of immunoassays and immunosensors by incorporating different kinds of nanostructures have gained considerable momentum over the last decade [13]. Labelling with nanoparticle is the core for an immunological diagnostic system by nanoparticle [14]. Luminescent quantum dot in immunoassay is the best example [15]. Basically, quantum dots are tiny light-emitting particles on the nanometer scale [16]. Quantum dots are emerging as a new class of biological labels with properties and applied usages that are not available with traditional organic dyes and fluorescent proteins [16]. At present, quantum dots are commonly used for diagnosis and imaging. For example, Zhu et al. recently reported quantum dots as a new immunofluorescent detection system for *Cryptosporidium parvum* and *Giardia lamblia* [17]. Zhu et al. said that this new fluorescence system exhibited superior photostability, gave 1.5- to 9-fold-higher signal-to-noise ratios than traditional organic dyes in detecting *C. parvum*, and allowed couple-color detection for *C. parvum* and *G. lamblia* [17]. More details on quantum dot are provided in another chapter in quantum medicine in this book.

In addition to quantum dots, there are applied usages of other nanomaterials for immunoassays. Superparamagnetic nanoparticles are a group that is commonly used at present. This technique is mentioned for its specific, sensitive, quantitative, and rapid detection of biological targets. In 2007, Kuma et al. reported their development of a liquid phase immunoassay system using magnetic nanoparticles [18]. According to their work, a highly sensitive immunoassay system using magnetic nanoparticles made from Fe3O4 was developed [18]. Kuma et al. said that since unbound magnetic markers randomly moved in solvent due to Brownian motion, there was no magnetic signal from unbound magnetic markers; therefore, the dividing of bound from unbound markers was not required [18]. In 2007, Wang and Tan reported the use of Fe3O4 magnetic nanoparticles/chitosan composite film for determination of ferritin [19]. They proposed that the analytical results revealed that the developed immunoassay was comparable with the radioimmunoassay, and the studied immunosensor exhibited good accuracy, high sensitivity, and long-term stability for three weeks, which implied a useful alternative approach for detecting ferritin in clinical diagnosis [19]. In 2005, Kim and Park described a new microfluidic immunoassay utilizing binding of superparamagnetic nanoparticles to beads and deflection of these beads in a magnetic field as the signal for measuring the existence of analyte [20]. According to this work, the velocities of microbeads conjugated with superparamagnetic nanoparticles were demonstrated by magnetic field gradients in microfluidic channels and compared with the calculated magnetic field gradients [20]. Kim and Park found that detection range and lower detection limit could be controlled by the microbead concentration and the higher magnetic field gradient [20]. In 2000, Chemla et al. reported an ultrasensitive magnetic biosensor for homogeneous immunoassay [21]. Their technique is a detection of biological targets by using superparamagnetic nanoparticles and a "microscope" based on a high-transition temperature dc superconducting quantum interference device (SQUID) [21]. In this way, a film to which the targets have been bound is placed on the microscope [21]. Chemla et al. found that nanoparticles that were bound to the target on the film were immobilized and undergo Néel relaxation, producing a slowly decaying magnetic flux, which is detected by the SQUID [21].

At present, the luminescent nanoparticle is an important nanomaterial type that is commonly used in nanodiagnosis. Luminescent silicon dioxide nanoparticle is a good example. The luminescent silicon dioxide nanoparticles with a size of 50 nm consisting of Rhodamine 6G can be synthesized by sol-gel method [22]. In the existence of Pb(Ac)2 as a heavy atom perturber, the particle can emit intense and stable room temperature phosphorescence signal of R, respectively, on polyamide membrane, with the lambda(ex)(max)/lambda(em)(max) = 470/635 nm for R [22]. In 2005, Jia-Ming et al. reported their work in producing a diagnostic test for determination of human IgG by solid substrate room temperature phosphorescence immunoassay based on an antibody labeled with nanoparticles consisting of Rhodamine 6G luminescent molecules [22]. A similar system was also completely developed by Liu et al. [23]. Recently, Liu et al. also reported their study on determination of alkaline phosphatase based on affinity adsorption solid-substrate room temperature phosphorimetry using rhodamine 6G-dibromoluciferin luminescent nanoparticle to label lectin and prediction of diseases [24]. This study revealed that AA-SS-RTP, whether by the sandwich method or the direct method, could combine very well the characteristics of both high sensitivity of solid-substrate room temperature phosphorimetry and specificity of the immunoreaction [24]. Europium-doped LaF3 nanoparticle is another group of luminescent nanomaterial [25]. Europium-doped LaF3 nanoparticles can be treated by the ionic reaction in the ethanol at 60 degrees C [25]. Meng et al. said that fluorescence intensity of the Europium-doped LaF3 nanoparticles varied remarkably with calcination temperatures [26]. Meng et al. noted that it was found that samples without any further calcinations can emit quite strong fluorescence [26]. Applied usage of Europium-doped LaF3 in nanodiagnosis is similar to that of nanoparticles consist ofing Rhodamine 6G luminescent molecules [26]. In 2006, Nichkova et al. reported for the first time the applied usage of europium-doped gadolinium oxide nanoparticles to the optical imaging of antibody micropatterns [27]. Nichkova et al. said that the nanoparticles could be used as fluorescent markers in a multiple appearance of immunosensing applied usages in a microscale format [27]. The cadmium telluride nanocrystal is another luminescent nanoparticle for nanodiagnosis. The favorable properties of hydro-soluble cadmium telluride nanocrystals as a new biological luminescent label over conventional fluorescent probes have attracted considerable interest [28]. This method may be applied in many types of antibody-antigen systems [28]. Recently, Yun et al. reported their success in using cadmium telluride quantum dots as a proton flux sensor and applying it to detect H9 avian influenza virus [29]. Finally, luminescent dye-doped silica nanoparticles or

FloDots have been developed for ultrasensitive bioanalysis and diagnosis in the past several years [30]. FloDots are highly luminescent and extremely photostable [30]. It is demonstrated that FloDots have many advantages over presently used luminescent probes, such as traditional fluorophores and quantum dots [30].

In addition to complex nanomaterials, simple nanomaterials such as the gold nanoparticle can also be used for immunodiagnosis. Wiwanitkit et al. recently reported the work in applied usage of gold nanoparticles for blood group testing [31]. Wiwanitkit et al. used the advanced nanomedicine method for modification of the basic blood group test for detection of weak B subgroup [31]. The preparation of gold nanoparticle solution in this study was according to the standard method [31]. In this work, Wiwanitkit et al. performed an experiment on many conditions on a weak B3 blood group sample, Anti B and gold nanoparticle solution [31]. In this study, Wiwanitkit et al. found that more agglutination could be detected in the system of antigen, antibody and gold nanoparticle than in the system only of antigen and antibody [31]. Wiwanitkit et al concluded that this applied usage could help diagnosis of weak blood group and could be the alternative method for the blood group test [31]. This is the first world report on this applied usage [31]. In 2005, Kato and Caruso also reported a homogeneous and competitive fluorescence quenching immunoassay based on gold nanoparticle/polyelectrolyte coated latex particles treated by the layer-by-layer method [32]. According to this study, nanoparticle/polyelectrolyte-coated latexes could be readily used as dynamic range tunable sensors [32]. In 2006, Hwang reported the use of a gold nanoparticle-based immunochromatographic test for identification of *Staphylococcus aureus* from clinical specimens [33]. Huang said that this method might be useful for analyzing *S. aureus* in a patient specimen [33]. Huang proposed that this method was rapid, easy to perform, and had a long shelf life at room temperature [33]. In 2006, Chen et al. also reported a new gold nanoparticles/sol-gel composite architecture for encapsulation of immunoconjugate for reagentless electrochemical immunoassay [34]. Chen et al. mentioned that this composite membrane could be used efficiently for the entrapment of different biomarkers and clinical applied usages [34].

NANOCANTILEVER

Nano-scale cantilever arrays are being developed for an ultra sensitive bio-assay [35]. Free-standing cantilevers, which directly translate specific biochemical reactions into micromechanical motion, have recently attracted much attention as label-free biosensors and micro/nano robotic devices [36]. In 2001, Fahlbusch and Fatikow integrated self-sensing SPM cantilevers into a gripper of microrobots [37]. Fahlbusch and Fatikow noted that these cantilevers run by measuring stress-induced electrical resistance changed in an implanted conductive channel in the flexure legs of the cantilever [37]. Fahlbusch and Fatikow said that the real-time force feedback delivered by these sensors enables us to better understand the prevailing nano forces and dynamics, which was indispensable for reliable micromanipulation strategies [37]. In 2003, Wang et al. performed an experiment on a nano-scale cantilver. In this work, the Zn-terminated ZnO polar surface was chemically active and the oxygen-terminated polar surface was inert in the growth of nanocantilever arrays [38]. According to this work, longer and wider "comblike" nanocantilever arrays were developed from the Zn surface, which was suggested to be a self-catalyzed process due to the enrichment of Zn at the growth front [38].

Molecular Diagnostic System by Nanoparticle

Molecular diagnosis is a commonly used technique in laboratory medicine at present. Over the past few years, progress in electrochemical biosensors for DNA analysis has been outstanding [39]. Applied usage of nanotechnology is the new advent. Using a multiple appearance of nanomaterials for molecular diagnostics will offer sensitive, rapid and cost-effective solutions for the novel clinical laboratory. There are many recent reports on molecular diagnostic systems by nanoparticles, as shown in Table 1.

Table 1. Recent reports on molecular diagnostic systems by nanoparticles

Authors	Details
Maki et al. [40]	Maki et al. reported nanowire-transistor based ultra-sensitive DNA methylation detection [40]. This work presented the first nanowire field effect transistor based biosensor technology which allows simple and ultra-sensitive electronic DNA methylation detection and avoids complicated bisulfite treatment and PCR amplification [40]. The producer of the p16(INK) gene, a tumor suppressor gene, is the target DNA in the detection model [40].
Kalogianni et al. [41]	Kalogianni et al. reported a simple and inexpensive assay that allowed visual detection and demonstration of the PCR-amplified sequences by hybridization within minutes [41]. In this work, the nanoparticles bound to the target DNA through hybridization, and the hybrids were captured by immobilized streptavidin at the test zone of the strip, producing a characteristic red line [41].
Wabuyele and Vo-Dinh [42]	Wabuyele and Vo-Dinh described the use of plasmonics-based nanoprobes that acted as molecular sentinels for DNA diagnostics [42]. In this work, the plasmonics nanoprobe was composed of a metal nanoparticle and a stem-loop DNA molecule tagged with a Raman label, and the nanoprobe utilized the specificity and selectivity of the DNA hairpin probe sequence to detect a specific target DNA sequence of interest [42]. Wabuyele and Vo-Dinh completely demonstrated the specificity and selectivity of the plasmonics nanoprobes to detect PCR amplicons of the HIV gene [42].
Fuentes et al. [43]	In this work, a single bond covalent immobilization of aminated DNA probes on magnetic particles suitable for selective molecular hybridization of traces of DNA samples was developed [43]. In this work, commercial superparamagnetic nanoparticles consisting of amino groups were activated by coating with a hetero-functional polymer (aldehyde-aspartic-dextran) [43]. According to this work, the experimental data supports the use of magnetic nanoparticles consisting of DNA probes immobilized by the procedure described as a convenient and extremely sensitive procedure for purification/detection of DNA/RNA from biological samples [43].
Shen et al. [44]	Shen et al. studied polymerase chain reaction (PCR) of nanoparticle-bound primers [44]. Shen et al. concluded that with either one or two primers respectively bound to the nanoparticle surface, PCR could proceed completely under optimized conditions and was subject to certain rules [44].

REFERENCES

[1] Emerich DF. Nanomedicine--prospective therapeutic and diagnostic applied usages. *Expert Opin. Biol. Ther.* 2005 Jan;5(1):1-5.

[2] 2 .Wang L, O'Donoghue MB, Tan W. Nanoparticles for multiplex diagnostics and imaging. *Nanomed.* 2006 Dec;1(4):413-26.

[3] Medvedeva NV, Ipatova OM, Ivanov IuD, Drozhzhin AI, Archakov AI. Nanobiotechnology and nanomedicine. *Biomed. Khim.* 2006 Nov-Dec;52(6):529-46.

[4] Maysinger D. Nanoparticles and cells: good companions and doomed partnerships. *Org. Biomol. Chem.* 2007 Aug 7;5(15):2335-42.

[5] Rotomskis R, Streckyte G, Karabanovas V. Nanoparticles in diagnostics and therapy: towards nanomedicine. *Medicina* (Kaunas). 2006;42(7):542-58.

[6] Service RF. Nanotechnology. Color-changing nanoparticles offer a golden ruler for molecules. *Science.* 2005;308:1099.

[7] Wiwanitkit V, Sereemaspun A, Rojanathanes R. Gold nanoparticle as an alternative tool for urine microalbumin test: the first world report. *Ren. Fail* (in press).

[8] Enüstün BV, Turkevich J. Coagulation of Colloidal Gold. *J. Am. Chem. Soc.* 1963; 85: 3317-3328.

[9] Tanaka T, Hayashi Y. Determination of silicon, calcium, magnesium and phosphorus in urine using inductively coupled plasma emission spectrometry and a matrix matching way. *Clin. Chim. Acta.* 1986; 156: 109-114.

[10] Domínguez-Vera JM, Colacio E. Nanoparticles of Prussian blue ferritin: a new route for obtaining nanomaterials. *Inorg. Chem.* 2003;42:6983-5.

[11] Wu ZS, Zhang SB, Guo MM, Chen CR, Shen GL, Yu RQ. Homogeneous, unmodified gold nanoparticle-based colorimetric assay of hydrogen peroxide. *Anal. Chim. Acta.* 2007 Feb. 12;584(1):122-8.

[12] Bauer M, Haglmuller J, Pittner F, Schalkhammer T. Direct optical visualization of DNA-DNA interaction by nanoparticle-capture on resonant PET-films. *J. Nanosci. Nanotechnol.* 2006 Dec;6(12):3671-6.

[13] Seydack M. Nanoparticle labels in immunosensing using optical detection methods. *Biosens. Bioelectron.* 2005 Jun 15;20(12):2454-69.

[14] Hempen C, Karst U. Tagging strategies for bioassays. *Anal. Bioanal. Chem.* 2006 Feb;384(3):572-83.

[15] Goldman ER, Medintz IL, Mattoussi H. Luminescent quantum dots in immunoassays. *Anal. Bioanal. Chem.* 2006 Feb;384(3):560-3.

[16] Smith AM, Nie S. Chemical analysis and cellular imaging with quantum dots. *Analyst.* 2004 Aug;129(8):672-7.

[17] Zhu L, Ang S, Liu WT. Quantum dots as a new immunofluorescent detection system for Cryptosporidium parvum and Giardia lamblia. *Appl. Environ. Microbiol.* 2004 Jan;70(1):597-8.

[18] Kuma H, Tsukamoto A, Saitoh K, Sugiura Y, Yoshinaga K, Enpuku K, Hamasaki N. Development of liquid phase immunoassay system using magnetic nanoparticles. *Rinsho Byori.* 2007 Apr;55(4):351-7.

[19] Wang SF, Tan YM. A new amperometric immunosensor based on Fe3O4 magnetic nanoparticles/chitosan composite film for determination of ferritin. *Anal. Bioanal. Chem.* 2007 Jan;387(2):703-8.

[20] Kim KS, Park JK. Magnetic force-based multiplexed immunoassay using superparamagnetic nanoparticles in microfluidic channel. *Lab. Chip.* 2005 Jun;5(6):657-64.

[21] Chemla YR, Grossman HL, Poon Y, McDermott R, Stevens R, Alper MD, Clarke J. Ultrasensitive magnetic biosensor for homogeneous immunoassay. *Proc. Natl. Acad. Sci. U S A.* 2000 Dec 19;97(26):14268-72.

[22] Jia-Ming L, Hui ZG, Aihong W, Pingping L, Huanhuan X, Li LD, Liu ZB. Determination of human IgG by solid substrate room temperature phosphorescence immunoassay based on an antibody labeled with nanoparticles consist ofing Rhodamine 6G luminescent molecules. *Spectrochim. Acta A Mol. Biomol. Spectrosc.* 2005 Mar;61(5):923-7.

[23] Liu JM, Yang TL, Liang XS, Wu AH, Li LD, Lin SQ. Determination of human IgG by solid-substrate room-temperature phosphorescence immunoassay based on an antibody labeled with nanoparticles consist ofing rhodamine 6G luminescent molecules. *Anal. Bioanal. Chem.* 2004 Oct;380(4):632-6.

[24] Liu JM, Liu ZB, Hu LX, He HX, Yang ML, Zhou P, Chen XH, Zheng MM, Zeng XY, Xu YL. Determination of alkaline phosphatase based on affinity adsorption solid-substrate room temperature phosphorimetry using rhodamine 6G-dibromoluciferin luminescent nanoparticle to label lectin and prediction of diseases. *Anal. Biochem.* 2006 Oct 15;357(2):173-80.

[25] Pi D, Wang F, Fan X, Wang M, Zhang Y. Luminescence property of Eu3+ doped LaF3 nanoparticles. *Spectrochim. Acta A Mol. Biomol. Spectrosc.* 2005 Sep;61(11-12):2455-9.

[26] Meng JX, Zhang MF, Liu YL, Man SQ. Hydrothermal preparation and luminescence of LaF3:Eu3+ nanoparticles. *Spectrochim Acta A Mol. Biomol. Spectrosc.* 2007 Jan;66(1):81-5.

[27] Nichkova M, Dosev D, Perron R, Gee SJ, Hammock BD, Kennedy IM. Eu3+-doped Gd2O3 nanoparticles as reporters for optical detection and visualization of antibodies patterned by microcontact printing. *Anal. Bioanal. Chem.* 2006 Feb;384(3):631-7.

[28] Li X, Wang L, Zhou C, Guan T, Li J, Zhang Y. Preliminary studies of applied usage of CdTe nanocrystals and dextran-Fe3O4 magnetic nanoparticles in sandwich immunoassay. *Clin. Chim. Acta.* 2007 Mar;378(1-2):168-74.

[29] Yun Z, Zhengtao D, Jiachang Y, Fangqiong T, Qun W. Using cadmium telluride quantum dots as a proton flux sensor and applying to detect H9 avian influenza virus. *Anal. Biochem.* 2007 May 15;364(2):122-7.

[30] Yao G, Wang L, Wu Y, Smith J, Xu J, Zhao W, Lee E, Tan W. FloDots: luminescent nanoparticles. *Anal. Bioanal. Chem.* 2006 Jun;385(3):518-24.

[31] Wiwanitkit V, Seereemaspun A, Rojanathanes R. Increasing the agglutination reaction in slide test for weak B blood group by gold nanoparticle solution: The first world report. *J. Immunol. Methods.* 2007 Oct 4;

[32] Kato N, Caruso F. Homogeneous, competitive fluorescence quenching immunoassay based on gold nanoparticle/polyelectrolyte coated latex particles. *J. Phys. Chem. B.* 2005 Oct 27;109(42):19604-12.

[33] Hwang SH. Gold nanoparticle-based immunochromatographic test for identification of Staphylococcus aureus from clinical specimens. *Clin. Chim. Acta.* 2006 Nov;373(1-2):139-43.

[34] Chen J, Tang J, Yan F, Ju H. A gold nanoparticles/sol-gel composite architecture for encapsulation of immunoconjugate for reagentless electrochemical immunoassay. *Biomaterials*. 2006 Apr;27(10):2313-21.

[35] Lin H, Datar RH. Medical applied usages of nanotechnology. *Natl. Med. J. India.* 2006 Jan-Feb;19(1):27-32.

[36] Fahlbusch S, Fatikow S. Implementation of self-sensing SPM cantilevers for nano-force measurement in microrobotics. *Ultramicroscopy*. 2001 Jan;86(1-2):181-90.

[37] Wang ZL, Kong XY, Zuo JM. Induced growth of asymmetric nanocantilever arrays on polar surfaces. *Phys. Rev. Lett.* 2003 Oct 31;91(18):185502.

[38] Watari M, Galbraith J, Lang HP, Sousa M, Hegner M, Gerber C, Horton MA, McKendry RA. Investigating the molecular mechanisms of in-plane mechanochemistry on cantilever arrays. *J. Am. Chem. Soc*. 2007 Jan 24;129(3):601-9.

[39] Ju H, Zhao H. Electrochemical biosensors for DNA analysis. Front *Biosci.* 2005 Jan 1;10:37-46.

[40] Maki WC, Mishra NN, Cameron EG, Filanoski B, Rastogi SK, Maki GK. Nanowire-transistor based ultra-sensitive DNA methylation detection. *Biosens Bioelectron.* 2007 Aug 30;

[41] Kalogianni DP, Goura S, Aletras AJ, Christopoulos TK, Chanos MG, Christofidou M, Skoutelis A, Ioannou PC, Panagiotopoulos E. Dry reagent dipstick test combined with 23S rRNA PCR for molecular diagnosis of bacterial infection in arthroplasty. *Anal. Biochem.* 2007 Feb 15;361(2):169-75.

[42] Wabuyele MB, Vo-Dinh T. Detection of human immunodeficiency virus type 1 DNA sequence using plasmonics nanoprobes. *Anal. Chem.* 2005 Dec 1;77(23):7810-5.

[43] Fuentes M, Mateo C, Rodriguez A, Casqueiro M, Tercero JC, Riese HH, Fernandez-Lafuente R, Guisan JM. Detecting minimal traces of DNA using DNA covalently attached to superparamagnetic nanoparticles and direct PCR-ELISA. *Biosens. Bioelectron.* 2006 Feb 15;21(8):1574-80.

[44] Shen HB, Hu M, Wang YB, Zhou HQ. Polymerase chain reaction of nanoparticle-bound primers. *Biophys. Chem.* 2005 May 1;115(1):63-6.

Chapter 8

NANOPHARMACOLOGY

INTRODUCTION TO NANOPHARMACOLOGY

Steady progress in the identification of human pharmacogenetic variants and new discoveries of disease susceptibility genes makes the old notion of one disease/one medicine untenable [1]. The present concept of pharmacology is highly selective pharmacological therapy. This concept has been proposed for centuries. This concept was first propagated by Paul Ehrlich. The best example of the magic bullet concept is applies to cancer therapy [2- 3]. There is a great need for more specific targeting of chemotherapeutic agents, but development of specific therapy will be difficult in light of the barriers which separate a tumor from the vasculature, tumor cell heterogeneity and instability, technological advances necessary for medicine delivery design and introduction into the clinic and mechanisms for assessment of efficacy [4]. However, Kreuter et al. noted that behavioral and social strategies will not come in the form of pre-packaged, easily exported "magic bullets," complete with efficacy estimates, for the prevention and control of selected cancers [5].

Nanotechnology is the new hope to lead us directly to the magic bullet [6 – 8]. Indeed, the technology is expected to create innovations and play a critical role in various biomedical applied usages, not only in medicine delivery, but also in molecular imaging, biomarkers and biosensors [9]. Target-specific medicine therapy and methods for early diagnosis of pathologies are the priority research areas in which nanotechnology would play an important role [9]. During the past 30 years, the explosive growth of nanotechnology has burst into challenging innovations in pharmacology, the main input being the ability to perform temporal and spatial site-specific delivery [6 – 8]. This has led to some marketed compounds during the last decade [6].

NANOTECHNOLOGY AND MEDICINE DESIGN

Introducing foreign bodies into the complex machinery of the human body is, however, a great and humbling challenge [10]. In order for nanobiology to reach its full potential, a means to alter the properties of nanoparticles, as expressed in the human body, in a predictable manner is needed [10]. Computer-aided methods are the natural option to speed up the development of these technologies [11]. In silico nanobio-design is an important method. The important step for in silico medicine design will be further described.

A. Genome Sequence Analysis

Genome sequence analysis is useful in the search for new medicines, especially for infectious diseases. The availability of a complete microbial genome sequence in 1995 brought the start of a genomic era that has allowed medical scientists to change the paradigm and approach vaccine development starting from genomic information [12]. The whole-genome perspective is expected to bring a contribution to medicine and vaccine development, particularly to target those germs for which the traditional approaches have failed so far [13]. Combining germ genome sequences with the host and vector genome sequences is useful to be a robust method for the identification of host-germ interactions. In addition, comparative sequencing of related species, especially of organisms used as model systems in the study of the disease, is beginning to realize its potential in the identification of genes that are involved in evasion of the host immune response [14]. These data are all useful in medicine and vaccine development [15].

B. Microarrays

Identification of an immune response correlate for protection against germs would greatly help the rational development of an effective vaccine. However, finding such a correlate has been a daunting task. DNA microarray technology is a new and powerful tool that allows the simultaneous analysis of a large number of nucleic acid hybridization experiments in a rapid and efficient fashion [16]. An advantage of microarray technology is that it can assist researchers to better define and understand the expression profile of a given genotype associated with disease, adverse effects from exposure to certain stimuli, or the ability to understand or predict immune responses to specific antigens [16]. The development of DNA microarray technology a decade ago brought the establishment of functional genomics as one of the most active and successful scientific disciplines at present [17].

C. Proteomics Approach and High-Throughput Cloning

Despite the increasing availability of genome sequences from many human genes, the production of complete proteomes remains at a bottleneck. Traditionally, the production of a recombinant protein requires a preliminary cloning step of the target gene into an expression vector before evaluating its cellular expression [18 – 19]. Among present methods, site-specific recombination cloning techniques, which eliminate the use of restriction endonucleases and ligase, offer usefulness in the context of high throughput procedures [19]. Rapid and highly efficient, recombinational cloning technology is largely used for structural genomics and functional proteomics [18 – 19]. High-throughput approaches for gene cloning and expression require the development of new, nonstandard tools for use by molecular biologists and biochemists [20]. High-throughput PCR recombination cloning and expression platform has been developed that allows hundreds of genes to be batch-processed by using basic laboratory procedures [21]. The derived data can be useful for medicine and vaccine development.

D. Bioinformatics Database Tools and Computational Vaccinology

Many advance bioinformatics have been launched for a few years and these database tools are useful for vaccine developments. Huge summations of data are produced by genomics and proteomics projects and large-scale screening of germ-host and antigen-host interactions. For vaccine discovery, one can "mine" the genomic sequence for potential surface targets using various algorithms, characterize these gene targets, and produce primers for cloning, all before one enters the wet laboratory [22]. Both the predictions and the interpretation of experiments rely on existing information in the literature which is mostly manually extracted and curated [23]. These can be addressed well by quantitative experimental and theoretical biophysical techniques, and particularly by methods from medicine design [24]. Databases and data mining are the two principal weapons at the disposal of in silico medicine search. Future development will also include systemic models of medicine and vaccine responses [25]. The common in silico methods include databases, quantitative structure-activity relationships, similarity searching, pharmacophores, homology models and other molecular modeling, machine learning, data mining, network analysis tools and data analysis tools that use a computer [26 – 31]. Such methods have seen frequent use in the discovery and optimization of new molecules with affinity to a target, the clarification of absorption, distribution, metabolism, excretion and toxicity properties, as well as physicochemical characterization [26 – 31]. More details can be seen in the chapter on nanoinformatics.

Jain said that nanotechnology can help medicine research to reach the state of personalized medicine, which implies a magic bullet [32]. Morrow et al. said that site-specific targeted medicine delivery and personalized medicine were just a few concepts that were on the horizon [33]. In conclusion, nanotechnology is proven to be extremely important for future personalized medicine and systems biology [34 – 35]. Indeed, nanotechnology realizes the usefulness of naturally occurring biological macromolecules and their building-block nature for design [36]. Frequently, assembly starts with the choice of a good molecule that is synthetically optimized towards the desired shape from the previously described nanobio-design [36]. However, Tsai et al. proposed another alternative, starting with a pre-specified nanostructure shape, selecting candidate protein building blocks from a library and mapping them onto the shape and, finally, testing the stability of the construct [36].

Nanotechnology and Medicine Delivery

In addition to medicine design, medicine delivery is another focus of interest [37 – 43]. Applied usage of nanotechnology for medicine delivery is accepted in pharmacology. At present, nanoparticles are considered to have the potential to be new intravascular or cellular probes for both diagnostic (imaging) and therapeutic purposes (medicine/gene delivery), which is expected to generate innovations and play a critical role in medicine [39]. The application of nanotechnology to medicine delivery can be seen in four important areas – controlled release of medicine, prolongation of medicine life-time, acceleration of medicine absorption, and medicine targeting.

A. Controlled Release of Medicine

To convey a sufficient dose of medicine to the lesion, suitable carriers of medicines are necessary [44]. Conventional dosage routes, such as oral delivery or injection, are the predominant routes for medicine delivery [45]. The main drawbacks of these types of dosages consist of essentially a limited control of the medicine-delivery rate and of the target area [45]. Nanoparticle carriers have important potential applications for the administration of therapeutic molecules without the quoted problem [44]. Nanoparticles are important controlled medicine-delivery devices [44]. There is much research in this area. Some interesting reports are presented in Table 1.

Table 1. Some interesting reports on nanotechnology for controlled release of medicines

Authors	Details
Ai et al. [46]	In this work, furosemide microcrystals were encapsulated with polyions and gelatin to control the release of the medicine in aqueous solutions [46]. Ai et al. noted that the results demonstrated a method of achieving prolonged medicine release through self-assembly of polymeric shells on medicine microcrystals [46].
Liu et al. [47]	Liu et al. performed a study of bovine serum albumin (BSA)-loaded calcium-deficient hydroxyapatite nano-carriers for controlled medicine delivery [47]. Liu et al. concluded that an initial bursting release was due to the desorption of BSA from the calcium-deficient hydroxyapatite surface, followed by a slow release as a result of the dissolution of the BSA-integrated nano-crystals along its c-axis direction [47].
Watanabe et al. [48]	In this work, two types of biocompatible nanoparticles with an average diameter of around 200 nm were formed only by mixing hydrolysates of chitosan and carboxymethyl cellulose; nanoparticle A was produced from chitosanase hydrolysate of chitosan and cellulase hydrolysate of carboxymethyl cellulose, and nanoparticle B was produced from lysozyme hydrolysate of chitosan and the carboxymethyl cellulose hydrolysate [48]. Watanabe et al. indicated that nanoparticle A could be applied to controlled-release medicine delivery systems, and that nanoparticle B was stably retained in the body without releasing the entrapped compounds [48].
Kang et al. [49]	The main objective of this study was to develop an optimal paclitaxel microemulsion treated by a self-microemulsifying medicine delivery system which was a mixture of paclitaxel, tetraglycol, Cremophor ELP, and Labrafil 1944 and a paclitaxel microemulsion consisting of poly(D,L-lactide-co-glycolide) in order to offer controlled release of paclitaxel [49].
Chen and Lin [50]	In this work, indomethacin micro-crystals approximately 2 microns in size were encapsulated with polyelectrolyte multi-layers for the purpose of controlled release [50]. Chen and Lin said that the results deomonstrated a method of achieving prolonged medicine release through self-assembly of polymeric shells on medicine microcrystals [50].

Authors	Details
Wei et al. [51]	This study reported an approach to fabricate tissue engineering scaffolds capable of controlled growth factor delivery whereby growth factor consisting of microshperical objects were integrated into three-dimensional scaffolds with good mechanical properties, well-interconnected macroporous and nano-fibrous structures [51]. Wei et al. concluded that successful generation of three-dimensional nano-fibrous scaffold incorporating controlled-release factors indicates significant potential for more complex tissue regeneration [51].
Jung et al. [52]	Jung et al. reported on the development of laponite-based nanohybrid for amplified solubility and controlled release of itraconazole [52]. In this work, laponite, a form of layered aluminosilicates, and itraconazole, a hydro insoluble medicine, were hybridized through an interfacial reaction at the boundary between hydro and a hydro-immiscible liquid [52].
Sammartino et al. [53]	In this report, Sammartino et al. presented the formulation, preparation and characterization of new polymeric composite materials consisting of a nano-hybrid to be used for the controlled molecular delivery of an anti-inflammatory molecule, Diclofenac [53].

B. Prolongation of Medicine Duration

Extension of drug duration is another applied usage of nanomaterial in medicine delivery. For example, the modification of medicines with hydro-soluble nanopolymer can help hold the medicine in the vessel. Zhou et al. mentioned that biodegradable colloidal nano-micelles were a new targeting medicine delivery and controlled release system, which could prolong the biological half-life and lighten the toxicity of chemotherapeutant while presenting fine biocompatibility [54]. In 2006, Pattani et al. reported work on development and comparative anti-microbial evaluation of lipid nanoparticles and nanoemulsion of Polymyxin B [55]. Pattani et al. reported that the developed lipid nanoparticles and nanoemulsion were useful delivery vectors for the anti-microbial medicines, and lipid nanoparticles could give an initial too as sustained effect while the nanoemulsion was capable of exerting a potent effect for a shorter period of time [55]. In 2007, Liu et al. reported another work on the fate of unimers and micelles of a poly(ethylene glycol)-block-poly(caprolactone) copolymer in mice following intravenous administration [56]. Liu et al. reported that the hydrophobic and semi-crystalline nature of the poly(ethylene glycol)-b-poly(caprolactone) core imparted a high degree of kinetic stability to this micelle system [56]. In 2004, Li et al. performed another interesting study to investigate the preparation technique, shape characteristics, and medicine releasing characteristics in vivo of 5-FU loaded core-shell type nanoparticles which were made by biodegradable amphiphlic poly (ethylene glycol)-poly (gamma-benzyl-L-glutamate)(PEG-PBLG) [57]. Li et al. also found that the 5-FU/PEG-PBLG nano-micelles could significantly affect the pharmacokinetics of 5-FU, prolonging the circulation time in vivo and had a slow release [57].

C. Acceleration of Medicine Absorption

Nanoparticles can help acceleration of medicine absorption by decreasing biological barriers of cells and tissues. Cevc said that colloids from an aqueous suspension could cross the skin barrier only through hydrophilic pathways [58]. Cevc also proposed that various colloids had a different ability to do this by penetrating narrow pores of fixed size in the skin, or the relevant nano-pores in barriers modeling the skin [58]. Liposome is a commonly used nanoparticle for acceleration of medicine absorption. Fang said that classic liposomes were of little value as carriers for medicine delivery via the skin because they did not penetrate it deeply [59]. Fang also mentioned that only specially designed liposomes had been shown to be capable of achieving amplified delivery [59]. Basically, manipulation of the physical properties of a medicine delivery system brings improved control over the pharmacokinetics and pharmacodynamics of the encapsulated medicines relative to free medicines [60]. Liposomes can act as a sustained release delivery system, and manipulation of properties such as liposome diameter, medicine release rate, bioavailability, and dosing schedule can significantly impact the therapeutic outcome of the liposomal medicines [60].

D. Medicine Targeting

Nanoparticles can help medicine targeting via medicine target recognition. Over the last few years, nano-structures have been developed in order to improve the efficiency and the specificity of medicine action [61]. Their small size allows them to be injected intravenously in order to reach target tissues [61]. This applied usage is previously discussed in the concept of magic bulleting. It is commonly used for anticancer medicines. Selective delivery to solid tumor sites can be done by utilizing amplified permeability and retention effect of solid tumor sites [62]. Essential factors of carriers systems based on this mechanism size and chemical character of carriers [62]. To design an appropriate nanoparticle that is most suitable for medicine targeting, layer-by-layer assembly can be used as a bottom-up nanofabrication way [63]. There are many nanomaterials that are presently applied for medicine delivery. However, the most commonly used ones are aptamer, liposome, and dendrimer.

E. Aptamer

Numerous nucleic acid ligands, also termed decoys or aptamers, have been developed during the past 15 years that can inhibit the activity of many pathogenic proteins [64]. Aptamers may prove useful in the treatment of a wide range of human disorders, including infectious diseases, cancer, and cardiovascular disease. Some interesting reports on this subject are presented in Table 2.

Table 2. Some interesting reports on aptamer for medicine delivery

Authors	Details
Farokhzad et al. [65]	Farokhzad et al. synthesized poly(lactic acid)-block-polyethylene glycol (PEG) copolymer with a terminal carboxylic acid functional group (PLA-PEG-COOH), and encapsulated rhodamine-labeled dextran (as a model medicine) within PLA-PEG-COOH nanoparticles [65]. According to this work, Farokhzad et al .demonstrated that these bioconjugates could efficiently target and be taken up by the prostate LNCaP epithelial cells, which expressed the prostate-specific membrane antigen protein [65].
Bagalkot et al. [66]	Bagalkot et al. reported a new quantum dot - aptamer-doxorubicin conjugate as a targeted cancer imaging, therapy, and sensing system [66]. Bagalkot et al. demonstrated favorable specificity and sensitivity of this nanoparticle conjugate as a cancer imaging, therapy and sensing system in vitro [66].
Kanopka et al. [67]	Kanopka et al. examined whether HIV-1 gene expression could be inhibited by the anti-HIV Rev-binding aptamer, and whether the antiviral effect of the aptamer could be amplified by a ribozyme directed against the HIV-1 env gene [67]. Kanopka et al demonstrated strong support for the use of in vitro evolved ligands as potential anti-HIV agents [67].

F. Liposome

Liposome is a spherical shaped nanopolymer, and its main composition of liposome is sugar and protein that can act as the receptor (more details can be seen in another chapter in this book). This property of liposome can help accelerate medicine absorption as previously mentioned. In many cases, liposomal medicines are administered via the bloodstream [68]. The stability in the bloodstream, clearance, and biodistribution are dependent on the composition, size, and charge of the liposomes [68]. Rigid, small-size (100-200 nm) liposomes tend to be retained in the blood without degradation [68]. There have been many recent advents in liposome technology. Long-circulating liposome is a good example [68]. Three kinds of functional long-circulating liposomes are available at present, namely, thermosensitive liposomes for delivering macromolecules, pH-sensitive liposomes for the cytosolic delivery of encapsulated materials, and reticuloendothelial system – avoiding liposomes for the passive targeting of tumor tissues [69]. In addition, it has become increasingly evident that tissues utilize specific localization of enzymes to perform certain tasks, often associated with various types of tissue remodeling [70]. The ubiquitous existence of such enzymes, along with their specific localizations, presents an ideal opportunity to elicit specific delivery via an enzyme-triggered mechanism [70].

G. Dendrimer

Dendrimer is a synthesized nanopolymer with a dendrogram, three-branched style, structure (more details are presented in another chapter in this book). Dendrimers can significantly improve pharmacokinetic and pharmacodynamic properties of low molecular weight and protein-based therapeutic agents [71]. Furthermore, fluorescent antibodies and imaging contrast agents can be bound to these new polymers and the resulting complexes can be used for analyzing biological fluids and for diagnosis [71].

H. Transfersome

Transfersomes (Idea AG) are a form of elastic or deformable vesicle, which were first introduced in the early 1990s [72]. Elasticity can be generated by incorporation of an edge activator in the lipid bilayer structure [72]. The main composition of these vesicles is soya phosphatidyl choline incorporating sodium cholate and a small concentration of ethanol [72]. Transfersomes can be applied in a non-occluded method to the skin and have been shown to permeate through the stratum corneum lipid lamellar regions as a result of the hydration or osmotic force in the skin [72].

NANOVACCINOLOGY

Similar to nanopharmacology, computer-aided methods are the basic option to speed up the development in nanovaccinology [73]. Basically, vaccine development includes finding the epitope of targeted protein, producing a recombinant, and testing for its efficacy. Routinely, this process starts from in-vitro research to in-vivo research (from animal models to human clinical studies). In addition to the classical immunological method, with the advent of bioinformatics, development of a new cancer vaccine can be performed by advanced in silico techniques. The important steps for vaccine development will be further explained and discussed.

A. Finding Epitope

To find a peptide candidate for a vaccine, the starting point is to find proper epitopes. Different procedures for epitope mapping are available at present [74]. It should be noted that synthetic peptide vaccines aiming at the induction of a protective response against malignant diseases are commonly tested, but despite their success in animal models, they do not yet live up to their promise in humans [75]. Mapping for probable epitopes seems to be a hard step. The evidence accumulating from many recent studies points to a broader range of targets recognized than previously expected, in terms of both numbers and characteristics of the targeted antigens [76]. Also, multiple studies report a substantial variation in the targets recognized in different human individuals [76]. To solve the problem of heterogenicity, many new computational tools have been developed for finding probable candidate vaccine epitopes. Prediction of both T-cell and B-cell epitopes for further cancer vaccine development can be easily performed. Nanoinformatics can help these processes (more details are presented in the chapter on nanoinformatics in this book).

B. Development of New Vaccine Recombinant

Development of new vaccine recombinant from an alternative epitope is the next step. This can be done based on the principle of artificial genetic recombinant. To develop a new genetic recombinant, advanced gene delivery and transfer techniques can be used. Of the many methods, gene transfer by virus or transfection in the most commonly used. The nanoparticle can help gene delivery (more details are presented in another chapter in this book).

C. Testing for New Vaccine Efficacy

Testing for new vaccine efficacy is the final step of vaccine development. This can be performed as the described process under the heading of research into a cancer vaccine. Clinical experiments should be performed before widespread administration of the new vaccine. In order to decrease the long phase of clinical experiment, the application of advanced medical informatics technology, gene ontology [77], can help predict the function or phenotypic expression of a newly developed vaccine.

D. Adaptation of the Existing Vaccine Methods by Nanotechnology

In addition to searching for and production of a new vaccine, adaptation of the existing vaccine methods by nanotechnology can be performed. The adaptation makes use of nanoparticles mostly for delivery purposes. Some important reports on this subject can be seen in Table 3.

Table 3. Some interesting reports on adaptation of the existing vaccine techniques by nanotechnology

Authors	Details
Zhang et al. [78]	In this work, a new label-free amperometric immunosensor for the detection of rubella vaccine was developed by immobilizing anti-rubella serum on bilayer nano-Au/polymerized o-phenylenediamine film with electrodeposited [78].
Balenga et al. [79]	Balenga et al. assessd the use of a new gene porter (Den123--a nontoxic self-integrated dendritic spheroidal nanoparticle made of biodegradable monomers), aiming to enhance and improve the desired immune response for protection from allergy [79]. Balenga et al. concluded that administration of the DNA entrapped in Den123 nanoparticles caused sustained release of plasmids, Th1/Th2 balanced immune response with useful IgE inhibition [79].
Jain et al. [80]	Chitosan nanoparticles encapsulated vesicular systems for oral immunization was reported [80]. In this work, bovine serum albumin-loaded chitosan nanoparticles were treated and encapsulated in vesicles (liposomes and niosomes) to make them acid resistant upon oral administration [80].

REFERENCES

[1] Erickson RP. From "magic bullet" to "specially engineered shotgun loads": the new genetics and the need for indivicoupleized pharmacotherapy. *Bioessays.* 1998 Aug;20(8):683-5.

[2] Osieka R. Oncology. *Arzneimittelforschung.* 1997 Oct;47(10):1161-5.

[3] Galanski M, Keppler BK. Searching for the magic bullet: anticancer platinum medicines which can be accumulated or activated in the tumor tissue. *Anticancer Agents Med. Chem.* 2007 Jan;7(1):55-73.

[4] Welch DR. Biologic considerations for medicine targeting in cancer patients. *Cancer Treat Rev.* 1987 Dec;14(3-4):351-8.

[5] Kreuter MW. Human behavior and cancer. Forget the magic bullet! *Cancer.* 1993 Aug 1;72(3 Suppl):996-1001.

[6] Couvreur P, Vauthier C. Nanotechnology: intelligent design to treat complex disease. *Pharm. Res.* 2006 Jul;23(7):1417-50.

[7] Ullman D. Let's have a serious discussion of nanopharmacology and homeopathy. *FASEB J.* 2006 Dec;20(14):2661; author reply 2662. No

[8] Yokoyama M. Medicine targeting with nano-sized carrier systems. *J. Artif. Organs.* 2005;8(2):77-84.

[9] Sahoo SK, Labhasetwar V. Nanotech approaches to medicine delivery and imaging. *Medicine. Discov. Today.* 2003 Dec 15;8(24):1112-20.

[10] Cachau RE, Gonzalez-Nilo FD, Ventura ON, Fritts MJ. In-silico nanobio-design. A new frontier in computational biology. *Curr .Top Med. Chem.* 2007;7(15):1537-40.

[11] Loew GH, Villar HO, Alkorta I. Strategies for indirect computer-aided medicine design. *Pharm. Res.* 1993 Apr;10(4):475-86.

[12] Serruto, D., Rappuoli, R. Post-genomic vaccine development. *FEBS Lett.* 2006; 580: 2985-92.

[13] Scarselli, M., Giuliani, M.M., Adu-Bobie, J., Pizza, M., Rappuoli, R. The impact of genomics on vaccine design. *Potentials Biotechnol.* 2005; 23: 84-91.

[14] Capecchi, B., Serruto, D., Adu-Bobie, J., Rappuoli, R., Pizza, M. The genome revolution in vaccine research. *Curr. Issues Mol. Biol.* 2004; 6: 17-27.

[15] Wiwanitkit V.

[16] Dhiman, N., Bonilla, R., O'Kane, D,J., Poland, G.A. Gene expression microarrays: a 21st century tool for directed vaccine design. *Vaccine* 2001; 20: 22-30.

[17] Grandi, G. Rational antibacterial vaccine design through genomic technologies. *Int. J. Parasitol.* 2003; 35: 615 – 20.

[18] Barry, M.A., Howell, D.P., Andersson, H.A., Chen, J.L., Singh, R.A. Expression library immunization to discover and improve vaccine antigens. *Immunol. Rev.* 2004; 199; 68 – 83.

[19] Betton, J.M. Rapid and highly efficient, the recombinational cloning technology is largely used for structural genomics and functional proteomics. *Biochimie.* 2004; 86: :601-5.

[20] Dieckman, L.J., Hanly,W.C., Collart, E.R. Strategies for high-throughput gene cloning and expression. *Genet. Eng.* (N Y). 2006; 27: 179-90.

[21] Davies, D.H., Liang, X., Hernandez, J.E. et al Profiling the humoral immune response to infection by using proteome microarrays: high-throughput vaccine and diagnostic antigen discovery. *Proc. Natl. Acad. Sci. U S A*. 2005; 102: 547-52.

[22] Zagursky RJ, Russell D. Bioinformatics: use in bacterial vaccine discovery. *Bioways*. 2001; 31: 636, 638, 640.

[23] Schonbach, C., Nagashima, T., Konagaya, A. Textmining in support of knowledge discovery for vaccine development. *Methods*. 2004; 34: 488-95.

[24] Flower, D.R., McSparron, H., Blythe, M.J. et al Computational vaccinology: quantitative approaches. *Novartis Found Symp*. 2003; 254: 102-20.

[25] Brusic, V., August, J.T., Petrovsky, N. Information technologies for vaccine research. *Expert Rev. Vaccines*. 2005; 4: 407-17.

[26] Schuster D, Steindl TM, Langer T. Predicting medicine metabolism induction in silico. *Curr. Top Med. Chem*. 2006;6(15):1627-40.

[27] Ekins S, Mestres J, Testa B. In silico pharmacology for medicine discovery: applied usages to targets and beyond. B*r. J. Pharmacol.* 2007 Sep;152(1):21-37.

[28] Rollinger JM, Langer T, Stuppner H. Integrated in silico tools for exploiting the natural products' bioactivity. *Planta Med*. 2006 Jun;72(8):671-8.

[29] Thibaut U. Bioinformatics and rational medicine design: tools for discovery and better understanding of biological targets and mode of action of medicines. *Scand. J. Gastroenterol. Suppl*. 2002;(236):95-9.

[30] Hirono S. An introduction to the computer-aided structure-based medicine design--applied usages of bioinformatics to medicine discovery. *Rinsho Byori*. 2002 Jan;50(1):45-51.

[31] Fischer HP. Towards quantitative biology: integration of biological information to elucidate disease pathways and to guide medicine discovery. *Biotechnol. Annu. Rev.* 2005;11:1-68.

[32] Jain KK. Challenges of medicine discovery for personalized medicine. *Curr. Opin. Mol. Ther.* 2006 Dec;8(6):487-92.

[33] Morrow KJ Jr, Bawa R, Wei C. Recent advances in basic and clinical nanomedicine. *Med. Clin. North. Am*. 2007 Sep;91(5):805-43.

[34] Nie S, Xing Y, Kim GJ, Simons JW. PharmaMedDevice 2007. Issues in nanomedicine delivery and personalized medicine. *IMedicines*. 2007 Jul;10(7):455-8.

[35] Baba Y. Nanotechnology in medicine. *Nippon Rinsho*. 2006 Feb;64(2):189-98.

[36] Tsai CJ, Zheng J, Aleman C, Nussinov R. Structure by design: from single proteins and their building blocks to nanostructures. *Potentials Biotechnol*. 2006 Oct;24(10):449-54.

[37] Wei C, Wei W, Morris M, Kondo E, Gorbounov M, Tomalia DA. Nanomedicine and medicine delivery. *Med. Clin. North. Am*. 2007 Sep;91(5):863-70.

[38] Conti M, Tazzari V, Baccini C, Pertici G, Serino LP, De Giorgi U. Anticancer medicine delivery with nanoparticles. *In Vivo*. 2006 Nov-Dec;20(6A):697-701.

[39] Liu Y, Miyoshi H, Nakamura M. Nanomedicine for medicine delivery and imaging: a useful avenue for cancer therapy and diagnosis using targeted functional nanoparticles. *Int. J. Cancer*. 2007 Jun 15;120(12):2527-37.

[40] Shaffer C. Nanomedicine transforms medicine delivery. *Medicine Discov Today*. 2005 Dec;10(23-24):1581-2.

[41] Kim KY. Nanotechnology platforms and physiological challenges for cancer therapeutics. *Nanomedicine*. 2007 Jun;3(2):103-10.

[42] Rotomskis R, Streckyte G, Karabanovas V. Nanoparticles in diagnostics and therapy: towards nanomedicine. *Medicina* (Kaunas). 2006;42(7):542-58.

[43] Ashammakhi N. Nanosize, mega-impact, potential for medical applied usages of nanotechnology. *J. Craniofac. Surg.* 2006 Jan;17(1):3-7.

[44] Ravi Kumar MN. Nano and microparticles as controlled medicine delivery devices. *J. Pharm. Pharm Sci.* 2000 May-Aug;3(2):234-58.

[45] Pizzi M, De Martiis O, Grasso V. Fabrication of self integrated micro reservoirs for controlled medicine release. *Biomed. Microdevices*. 2004 Jun;6(2):155-8.

[46] Ai H, Jones SA, de Villiers MM, Lvov YM. Nano-encapsulation of furosemide microcrystals for controlled medicine release. *J. Control Release*. 2003 Jan 9;86(1):59-68.

[47] Liu TY, Chen SY, Liu DM, Liou SC. On the study of BSA-loaded calcium-deficient hydroxyapatite nano-carriers for controlled medicine delivery. *J. Control. Release*. 2005 Sep 20;107(1):112-21.

[48] Watanabe J, Iwamoto S, Ichikawa S. Entrapment of some compounds into biocompatible nano-sized particles and their releasing properties. *Colloids Surf. B Biointerfaces*. 2005 May 10;42(2):141-6.

[49] Kang BK, Chon SK, Kim SH, Jeong SY, Kim MS, Cho SH, Lee HB, Khang G. Controlled release of paclitaxel from microemulsion consist ofing PLGA and evaluation of anti-tumor activity in vitro and in vivo. *Int. J. Pharm.* 2004 Nov 22;286(1-2):147-56.

[50] Chen Y, Lin X. Studies on the medicine release properties of nano-encapsulated indomethacin microparticles. *J. Microencapsul.* 2005 Feb;22(1):47-55.

[51] Wei G, Jin Q, Giannobile WV, Ma PX. Nano-fibrous scaffold for controlled delivery of recombinant human PDGF-BB. *J. Control. Release*. 2006 May 1;112(1):103-10.

[52] Jung H, Kim HM, Choy YB, Hwang SJ, Choy JH. Laponite-based nanohybrid for amplified solubility and controlled release of itraconazole. *Int. J. Pharm.* 2007 Aug 17;

[53] Sammartino G, Marenzi G, Tammaro L, Bolognese A, Calignano A, Costantino U, Califano L, Mastrangelo F, Tete' S, Vittoria V. Anti-inflammatory medicine incorporation into polymeric nano-hybrids for local controlled release. *Int. J. Immunopathol. Pharmacol.* 2005 Jul-Sep;18(3 Suppl):55-62.

[54] Zhou JJ, Chen RF, Tang QB, Zhou QB, Lu HW, Wang J. Preparation of 5-fluorouracil encapsulated in amphiphilic polysaccharide nano-micelles and its killing effect on hepatocarcinoma cell line HepG2. *Ai Zheng*. 2006 Dec;25(12):1459-63.

[55] Pattani AS, Mandawgade SD, Patravale VB. Development and comparative anti-microbial evaluation of lipid nanoparticles and nanoemulsion of Polymyxin B. *J. Nanosci. Nanotechnol.* 2006 Sep-Oct;6(9-10):2986-90.

[56] Liu J, Zeng F, Allen C. In vivo fate of unimers and micelles of a poly(ethylene glycol)-block-poly(caprolactone) copolymer in mice following intravenous administration. *Eur. J. Pharm. Biopharm.* 2007 Mar;65(3):309-19.

[57] Liu J, Zeng F, Allen C. In vivo fate of unimers and micelles of a poly(ethylene glycol)-block-poly(caprolactone) copolymer in mice following intravenous administration. *Eur. J. Pharm. Biopharm.* 2007 Mar;65(3):309-19.

[58] Cevc G. Lipid vesicles and other colloids as medicine carriers on the skin. *Adv. Medicine Deliv. Rev.* 2004 Mar 27;56(5):675-711.

[59] Fang JY. Nano- or submicron-sized liposomes as carriers for medicine delivery. *Chang Gung Med. J.* 2006 Jul-Aug;29(4):358-62.

[60] Allen TM, Cheng WW, Hare JI, Laginha KM. Pharmacokinetics and pharmacodynamics of lipidic nano-particles in cancer. *Anticancer Agents Med. Chem.* 2006 Nov;6(6):513-23.

[61] Vonarbourg A, Passirani C, Saulnier P, Benoit JP. Parameters influencing the stealthiness of colloidal medicine delivery systems. *Biomaterials.* 2006 Aug;27(24):4356-73.

[62] Yokoyama M, Okano T. Targeting of anti-cancer medicines with nano-sized carrier system. *Nippon Rinsho.* 1998 Dec;56(12):3227-34.

[63] Ariga K, Hill JP, Ji Q. Layer-by-layer assembly as a bottom-up nanofabrication way for exploratory research and realistic applied usage. *Phys. Chem. Chem. Phys.* 2007 May 21;9(19):2319-40.

[64] Nimjee SM, Rusconi CP, Sullenger BA. Aptamers: an emerging class of therapeutics. *Annu. Rev. Med.* 2005;56:555-83.

[65] Farokhzad OC, Jon S, Khademhosseini A, Tran TN, Lavan DA, Langer R. Nanoparticle-aptamer bioconjugates: a new approach for targeting prostate cancer cells. *Cancer Res.* 2004 Nov 1;64(21):7668-72.

[66] Bagalkot V, Zhang L, Levy-Nissenbaum E, Jon S, Kantoff PW, Langer R, Farokhzad OC. Quantum dot-aptamer conjugates for synchronous cancer imaging, therapy, and sensing of medicine delivery based on bi-fluorescence resonance energy transfer. *Nano Lett.* 2007 Oct;7(10):3065-70.

[67] Konopka K, Duzgunes N, Rossi J, Lee NS. Receptor ligand-helpd cationic liposome delivery of anti-HIV-1 Rev-binding aptamer and ribozyme DNAs. *J. Medicine. Target.* 1998;5(4):247-59.

[68] Oku N, Namba Y. Long-circulating liposomes. *Crit. Rev. Ther. Medicine. Carrier Syst.* 1994;11(4):231-70.

[69] Oku N. Basic study and applied usage of functional liposomes. *Yakugaku Zasshi.* 1995 Jul;115(7):483-98.

[70] Meers P. Enzyme-activated targeting of liposomes. *Adv Medicine Deliv Rev.* 2001 Dec 31;53(3):265-72.

[71] Bai S, Thomas C, Rawat A, Ahsan F. Recent progress in dendrimer-based nanocarriers. *Crit. Rev. Ther. Medicine Carrier Syst.* 2006;23(6):437-95.

[72] Benson HA. Transfersomes for transdermal medicine delivery. *Expert. Opin. Medicine. Deliv.* 2006 Nov;3(6):727-37.

[73] Lee LA, Wang Q. Adaptations of nanoscale viruses and other protein cages for medical applied usages. *Nanomedicine.* 2006 Sep;2(3):137-49.

[74] Corradin G. Antigen processing and presentation. *Immunol Lett.* 1990 Aug;25(1-3):11-3.

[75] Bijker MS, Melief CJ, Offringa R, van der Burg SH. Design and development of synthetic peptide vaccines: past, present and future. *Expert Rev. Vaccines.* 2007 Aug;6(4):591-603.

[76] Sette A, Peters B. Immune epitope mapping in the post-genomic era: lessons for vaccine development. *Curr. Opin. Immunol.* 2007 Feb;19(1):106-10.

[77] Ontology annotation: mapping genomic regions to biological function. *Curr. Opin. Chem. Biol.* 2007 Feb;11(1):4-11.

[78] Zhang L, Yuan R, Huang X, Chai Y, Tang D, Cao S. A new label-free amperometric immunosenor for rubella vaccine. *Anal. Bioanal Chem.* 2005 Mar;381(5):1036-40.

[79] Balenga NA, Zahedifard F, Weiss R, Sarbolouki MN, Thalhamer J, Rafati S. Protective efficiency of dendrosomes as new nano-sized adjuvants for DNA vaccination against birch pollen allergy. *J. Biotechnol.* 2006 Jul 25;124(3):602-14.

[80] Jain S, Sharma RK, Vyas SP. Chitosan nanoparticles encapsulated vesicular systems for oral immunization: preparation, in-vitro and in-vivo characterization. *J. Pharm. Pharmacol.* 2006 Mar;58(3):303-10.

Chapter 9

NANOHEMATOLOGY

INTRODUCTION TO NANOHEMATOLOGY [1]

Today, nanomedicine has branched out in hundreds of different directions, each of them embodying the key insight that the ability to structure materials and devices at the molecular scale can bring enormous sudden benefits in the research and practice of medicine [2]. Due to the advancement in nanoscience in the present day, many nanomedicine and nanobiotechnology methods have been launched. Those methods are actually new advents in science and medicine. The basic principle of nanonscience is the different physical and chemical properties of the nanoparticle and its larger forms. Nanohematology is an advent medical science. Nanohematology focuses on hematology at the nano-level. At present, there is much ongoing development research for a new hematological analyzer based on nanofluidics and nanobiosensor principles. This new generation of hematological analyzer is expected to be the solution for point-of-care testing in hematology. In addition to an advanced analyzer, the simple application of nanoparticle solution can also be used for diagnosis in hematology. The naked eye detection system is the best example. Nanotoxicology focusing on hematological system is another interesting facet of hematology. There are many recent publications concerning the disturbance of blood cells responding to exposure to nanoparticles. The applied usage of nanopharmacology to make a magic bullet focusing on hematological malignancy is another area of nanohematology. It is accepted that a nanoparticle-tagged regimen is the new hope for treatment of many cancers. Finally, nanoinformatics is an advanced computational medicine for in silico study of hematological phenomena and disorders; for example, analysis of hemoglobin disorders by application of computational tools for nanohematology. In addition, interaction between nanomolecules can be simulated.

NEW NANOHEMATOLOGICAL ANALYZER

As previously mentioned, there are a number of ongoing development studies for a new hematological analyzer based on nanofluidics and nanobiosensor principles. Nanodevices, such as nanowires, nanotubes, and nanocantilevers, can be combined with nanoarrays and nanofluidics to create integrated and automated nanodetection platforms [3]. Accurate, fast, and affordable analysis of the cellular component of blood is of prime interest for medicine and research at present. Point-of-care testing is the present aim of clinical pathology [4 – 10]. Blood-on-a-chip is a new concept [11]. Precise control over the cell microenvironment during dividing procedures and the ability to scale down the analysis to very small volumes of blood are among the most attractive capabilities of the new approaches [11]. With the advance of microfabrication, on-chip blood count has become a target for miniaturization aiming at providing cost-effective, functional, capable point-of-care devices and systems that use less than 100 nL of blood sample and generate measurement results within minutes [12]. Some interesting reports on this subject are presented in Table 1.

Table 1. Reports on nanohematological analyzer

Authors	Details
Kang et al. [13]	Kang et al. used DC-Dielectrophoresis (DC-DEP), the induced motion of the dielectric particles in a spatially non-uniform DC electric field; it is applied to separate biological cells by size, combined with the electroosmotic flow, wherebymixed biological cells of a few to tens of micrometers difference in diameter can be continuously separated into different collecting wells [13]. Kang et al. said that for separating target cells of a specific size, all that is required is to modify the voltage outputs of the electrodes [13].
Tai [12]	Tai developed a method for two-part leukocyte differential using whole blood [12]. Unlike methods used in conventional blood analyzers, undiluted blood samples were stained with a nano molecular nucleic acid stain, acridine orange, in this work [12]. According to this work, lymphocytes and granulocytes emitted fluorescent light at different peak frequency after interaction with the dye due to the difference in cellular composition [12]. Using the undiluted sample greatly decreased the sample preparation procedure, and reduced the total measurement time, the reagent, and the waste volume [12]. Tai said that these benefits made it a practical method for implementation in microdevices and throughput of one thousand leukocytes per second was demonstrated, which means the leukocyte differential could be accomplished in a couple of seconds [12].
Bardell et al. [14]	The microcytometer described in this paper is designed to count and classify platelets, red blood cells, and various white cell populations by means of laminate-based microfluidic flow channels, and light-scattering optics [14]. Bardell et al. said that the new lab-on-a-chip-based tools used proprietary optically resolved chemical and cellular analysis microfluidics techniques to simplify the analysis process, getting rid of sample pre-processing and tool calibration [14].

In addition to an advanced analyzer, the simple application of a nanoparticle solution can also be used for diagnosis in hematology. The naked eye detection system is the best example. In 2007, Wiwanitkit et al. reported the first world report on increasing the agglutination reaction in a slide test for a weak B blood group by gold nanoparticle solution [15]. The preparation of gold nanoparticle solution in this study was implemented according to the standard method reported by Frenkel et al. [15]. The authors performed an experiment on many conditions on a weak B3 blood group sample, Anti B and gold nanoparticle solution [15]. In this study, the authors found that more agglutination could be detected in the system of antigen, antibody and gold nanoparticle than in the system of only antigen and antibody

[15]. Wiwanitkit et al. noted that this applied usage could help diagnosis of a weak blood group and could be the alternative method for the blood group test [15].

Nanohematotoxicology

Nanotoxicology focusing on a hematological system is another interesting facet of hematology. There are many recent publications concerning the disturbance of blood cells responding to exposure to nanoparticles. The effect of nanoparticles on each component of a hematological system will be hereby discussed.

A. Effect of Nanoparticle on Red Blood Cell

The effect of environmental contamination on red blood cells is well known. For example, Wiwanitkit recently demonstrated the effect of volatile hydrocarbon solvent on red blood cells [16]. However, the effect of nanoparticles has been poorly clarified. Concerning the hematological system, there is very limited data in this area. Unlike other cells, red blood cells do not have receptors on their surfaces for transporting particles across the cell membrane via the process known as endocytosis. However, Soler et al. [17] recently demonstrated the magnetic nanoparticle internalization by mouse red blood cells. Therefore, the possibility of the toxicity of the nanoparticle on red blood cells can be expected. Rothen-Rutishauser et al. [18] found that the surface charge and the material of the particles did not influence their entering. These results suggest that particles may penetrate the red blood cell membrane by a still unknown mechanism different from phagocytosis and endocytosis. Thrall [19] proposed the query of whether red blood cells were defenseless against smaller nanoparticles. Recently, Wiwanitkit et al. reported a visualization of a gold nanoparticle on a microscopic picture of a red blood cell and made a further implication for probable risk of nanoparticle exposure [20]. In this work, Wiwanitkit performed a study to demonstrate whether a gold nanoparticle could be detected inside red blood cell in a microscopic test [20]. This study was performed as an experimental study [20]. A mixture of gold nanoparticle solution and blood sample was treated and further analyzed [20]. According to this work, accumulation of gold nanoparticles in the red blood cell could be observed after mixing the blood sample with gold nanoparticle solution [20]. However, no significant destruction of the red cells could be seen [20]. In addition, the effect of gold nanoparticles on red blood cells could be detected and the implication for the probable chronic toxicity of the accumulated gold nanoparticles in red cells is increased [20].

B. Effect of Nanoparticle on White Blood Cell

There are some recent reports on the application of nanoparticles for diagnostic purposes on the lymphocyte, such as immunophenotyping [21 - 22]. Pittet et al. proposed that labeled cells could be detected by magnetic resonance imaging, fluorescence reflectance imaging, fluorescence-mediated tomography, confocal microscopy and flow cytometry, and could be purified based on their fluorescent or magnetic properties [23]. However, there is no evidence of the direct effect of gold nanoparticles on white blood cells. The exact mechanism of direct penetration of the gold nanoparticle into the cytoplasm of lymphocytes needs further study. Because the reported pore size in the lymphocyte membrane is about 4 nm x 2.5 nm, which is smaller than the diameter of ordinary gold nanoparticle, the authors believed that there might be some undiscovered mechanism to help direct penetration of gold nanoparticles into the lymphocytes.

C. Effect of Nanoparticle on Platelet

Concerning the hematological system, there is very limited data on the effect of the nanoparticle on a platelet. Ever-increasing use of engineered carbon nanoparticles in nanopharmacology for selective imaging, sensor or medicine delivery systems has increased the potential for blood platelet-nanoparticle interactions [24]. Radomski et al. reported that carbon nanoparticles and microparticles have the ability to activate platelets and enhance vascular thrombosis [24]. However, there is no report on gold nanoparticles. The author hereby performed a simple study to assess the effect of gold nanoparticles on platelets seen by microscopic examination. A single-donor fresh EDTA blood sample was donated and used for further laboratory analysis. The sample was collected and transferred according to the standard procedure of a medical laboratory. From microscopy, the platelet is estimated about 10/oil field (platelet count = 155,000/mL). Following the classical Turkevich citrate reduction method [25], 9 nanometer-sized AuNPs were established and AuNP concentration was treated at 44 ppm. The AuNPs can be stored in the dark under 4°C for over a month. For the study of the direct in vitro effect of gold nanoparticle on the blood sample, first, the mixture between equal summation (500 microliter) of gold nanoparticle solution and EDTA blood sample was treated. After being left for 15 minutes, the morphological changes of the red blood cell were studied by clinical microscopy under high-power field. All of the laboratory analysis was performed in a reference ISO 15189 accredited laboratory of the Department of Laboratory Medicine, Faculty of Medicine, Chulalongkorn University. According to this work, it was revealed that after mixing the blood sample with gold nanoparticle solution, platelet aggregration can be observed. According to microscopical examination, the platelet is estimated about 2/oil field. Here, the author demonstrated that the gold nanoparticle can induce platelet aggregation. Although recently proposed, the mechanism for induction of bionanoparticles that present triple helical CRP motifs of sufficient length on a suitable scaffold is important for stimulatation of platelets [26] bygold nanoparticle containing no peptide. The mechanism of induction has to be further studied to construct platelet substitutes that have hemostatic properties over a wide range of shear rates [27]. Of interest, Okamura et al. proposed that a mixed system of fibrinogen gamma-chain carboxy-terminal sequence HHLGGAKQAGDV (H12)- and recombinant hydro-soluble moiety of the platelet glycoprotein (rGPIbalpha)-conjugated nanoparticles was a more effective platelet substitute and had amplified platelet aggregation properties [27]. The finding in this study can be useful for development of new hemostatic medicines.

However, there are some limitations in this work. This is a very preliminary study that describes platelet aggregation induced by gold nanoparticles in EDTA blood. This might not be aggregation in the accepted sense because the latter relies on extracellular calcium. The methods used to define the end points might not be satisfactory. For a start, EDTA is not a suitable anticoagulant for these studies and will cause platelet aggregates in normal individuals. The idea is good but needs to be addressed using state-of-the-art technology. More studies on platelet function and physiology in the laboratory in all aspects of platelet function, including signal transduction, are recommended.

D. Effect of Nanoparticle on Coagulation System

The effect of nanoparticles on the coagulation system is poorly reported. In 1997, Sahli et al. reported the interactions of poly(lactic acid) and poly(lactic acid-co-ethylene oxide) nanoparticles with the plasma factors of the coagulation system [28]. In this paper, Sahli et al. reported that PLA suspensions stabilized by sodium cholate (PLA-Ch) interacted with thrombin, factor V and calcium ions and formation of complexes and aggregates was induced by addition of calcium ions to PLA-Ch suspensions in the existence or in the absence of plasma [28]. On the contrary, Sahli et al. also found that PLA-PEO suspensions were remarkably inert towards the coagulation factors and calcium ions, even when cholate was present [28]. To study the effect of gold nanoparticle on blood coagulation system, the author performed a similar work as previously mentioned in the session of effect of a nanoparticle on a platelet. The author found that there was no significant different on the prothrombin time and activated thromboplastin between samples treated with and without gold nanoparticle solution.

E. Effect of Nanoparticle on Bone Marrow

There is no direct report on the effect of nanoparticles on bone marrow.

However, there are some interesting reports on the nanoparticle in the marrow stem cell as presented in Table 2.

Table 2. Reports on the nanoparticle in the marrow stem cell

Authors	Details
Chen et al. [29]	Chen et al. used quantum dots conjugated with integrin antibodies and performed studies to quantitatively demonstrate changes in the integrin dynamics during osteogenic differentiation of human bone marrow derived progenitor cells (BMPCs) [29]. Chen et al. concluded that that the altered integrin dynamics was correlated with BMPC differentiation and that the integrin lateral mobility was restricted by direct links to microfilaments [29].
Garon et al. [30]	Garon et al. reported an observation on quantum dot tagging and tracking of human leukemic, bone marrow and cord blood cells [30]. Garon et al. said that quantum dots had the ability to bind to specific cells of interest, be taken up by a diverse range of hematologic cells, and followed through many divisions and through differentiation [30]. Garon et al. noted that these results established quantum dots as extremely useful molecular imaging tools for the study of hematologic cells [30].

Table 2. Continued

Authors	Details
Horak et al.. [31]	Horak et al. reported a study on D-mannose-modified iron oxide nanoparticles for stem cell tagging [31]. In this work, new surface-modified iron oxide nanoparticles were developed by precipitation of Fe(II) and Fe(III) salts with ammonium hydroxide by two different methods [31]. In the first method, precipitation was done in the existence of D-mannose solution (in situ coating); the second method involved oxidation of precipitated magnetite with sodium hypochlorite followed by addition of D-mannose solution (postsynthesis coating) [31]. Horak et al. found that relaxivity measurements of labeled cells in gelatin revealed very high relaxivities only for postsynthesis D-mannose-coated iron oxide nanoparticles [31].
Zheng et al [32]	Zheng et al. reported their study on cellular imaging and surface marker tagging of hematopoietic cells using quantum dot bioconjugates [32]. The results of this study demonstrated the feasibility of using quantum dots for cell tagging and surface marker analysis of hematopoietic cells [32]. Zheng et al. said that given the superior optical properties of qdots as compared to conventional fluorescence dyes, the quantum dot-based probe offered a useful tool for hematology analysis in clinical laboratories [32].

NANOHEMATOONCOLOGY

The applied usage of nanopharmacology to make a magic bullet focusing on hematological malignancy is another area of nanohematology. It is accepted that nanoparticle-tagged regimen is the new hope for treatment of many cancers. Nanoparticular pharmaceutical or medical products are used for new therapeutic approaches, which are able to enter specifically certain types of tumor tissues or tumor cells, so that the local effectiveness of former systemic therapies may be improved [33]. Portney and Ozkan said that innovation in the last decade has endowed nanotechnology with an assortment of tools for delivery, imaging, and sensing in cancer research – stealthy nanoparticle vectors circulating in vivo, integrated with exquisite molecular control, capable of selective tumor targeting and potent delivery of therapeutics; intense and photostable quantum dot-based tumor imaging, enabling multicolor detection of cell receptors with a single optical excitation source; arrays of semiconducting nanowire and carbon nanotube sensor elements for selective multiplexed sensing of cancer markers without the need for probe tagging [34]. Maysinger noted that nanooncology should be discussed in more detail in the context of the intracellular fate of nanoparticles and probable long-term consequences on cell functions [35]. Some interesting reports on the nanohematooncology are presented in Table 3 and Table 4.

Table 3. Reports on the nanohematooncology (solid tumor aspect)

Authors	Details
Choi et al. [36]	Choi et al. reported that quantum dot core and surface compositions, as well as quantum dot stability, all influenced nanoparticle internalization and the consequent cytotoxicity [36]. Choi et al. also reported that cadmium telluride quantum dot-induced toxicity involved the upregulation of the Fas receptor and lipid peroxidation, leading to impaired neuroblastoma cell functions [36]. Choi et al noted that further improvements of nanoparticles and scientists' understanding of the underlying mechanisms of quantum-toxicity were critical for the development of new nanotherapeutics or diagnostics in nanooncology [36].
Schatzlein et al. [37]	Schatzlein et al. exploited fd phage-derived peptides for the targeting of low molecular weight medicine conjugates to solid tumours [37]. As a model, Schatzlein et al. chose doxorubicin conjugates targeted to the transferrin receptor [37]. According to this work, the doxorubicin conjugate of the peptide consist ofing a cleavable linker was treated and endosomal uptake warranted by fluorescence microscopy [37].
He et al. [38]	He et al. reported changes induced by adenovirus-delivered wild-type p53 gene and chemotherapy of U87 MG glioblastoma cells, a treatment known to trigger apoptosis and cell cycle arrest [38]. In this work, He et al. compared polar lipid changes in treated cells and control cells by use of a new, sensitive method that employs lipid extraction, one-step liquid chromatography dividing, high-resolution mass analysis, and Kendrick mass defect analysis [38]. He et al. found that sulfatides were most highly modulated by wild-type p53 treatment and the treatment also led to an increase in phospholipids such as phosphatidyl inositols, phosphatidyl serines, phosphatidyl glycerols, and phosphatidyl ethanolamines [38].

Table 4. Some interesting reports on the nanohematooncology (leukemia)

Authors	Details
Li et al. [39]	According to this work, influence of a discrete nano-hydroxyapatite crystal (nano-HAp) on lymphatic leukemia P388 cell property was investigated by an in vitro way using an MTT test and FITC analysis [39]. Li et al. proved that the discrete nano-HAp could cause apoptosis of the leukemia P388 cells, and selectively act on G(1) phase and arrest the G(1) phase in cell cycle of the leukemia P388 [39].
Luo et al. [40]	Luo et al. performed this study to investigate the proliferation inhibition and the differentiation effects of realgar (As4S4) nano-particles on human acute myeloid leukemia cell line HL-60 [40]. Luo et al concluded that low dose realgar induced differentiation in human acute myeloid leukemia cell line HL-60 [40].
Lapotko et al. [41]	In this work, experimental investigation of laser nano-thermolysis of leukemia cells was performed using 30 nm spherical gold nanoparticles as a light absorbing agent, and photothermal and fluorescent microscopies, too, as flow cytometry as methods to monitor microbubble formation and resulting damage of leukemia cells in human bone marrow specimens [41]. Lapotko et al. concluded that two-stage cell targeting method allows formation of nanoparticle clusters selectively in diagnosis-specific tumor cells and the clusters served as effective sources of photothermally-induced microbubbles, which killed indivicouple target cells after a single laser pulse [41]. Lapotko et al. also mentioned that the laser fluence threshold for generation of microbubbles was inversely proportional to the volume of nanoparticle clusters [41].

Table 4. (Continued)

Authors	Details
Xia et al. [42]	Xia et al. performed a study to identify two differentiation-associated proteins induced by rhIL-6 in M1 mouse myeloid leukemia cells [42]. Xia et al. mentioned that nano-electrospray ionization mass spectrometry/mass spectrometry way could completely identify the two differentiation-associated proteins induced by rhIL-6 and had great usefulness in protein analysis [42].
Rezaul et al. [43]	Rezaul et al. reported for a systematic characterization of mitochondrial proteome from human T leukemia cells [43]. Rezaul et al. explored the feasibility and efficiency of protein identification using the one-dimensional gel electrophoresis in combination with the nano liquid-chromatography tandem mass spectrometry [43]. Rezaul et al. mentioned that this approach was suitable for characterizing subcellular proteomes from multiple cells and tissues [43].
Ning et al. [44]	Ning et al. performed a study to test the growth-inhibitory, apoptosis- and necrosis-inducing effects of realgar nanoparticles in human chronic myelogenous leukemia cell line K562 and acute myeloid leukemia cell line HL-60, and to find out the chemical species with efficacy [44]. Ning et al. concluded that realgar nanoparticle induced both apoptosis and necrosis in leukemia cell lines K562 and HL-60. Thioarsenite species with both As-O and As-S bonds may be the active intermediates in the realgar nanoparticle.

APPLIED NANOINFORMATICS FOR HEMATOLOGY

Nanoinformatics, a new branch of bioinformatics, is an advanced computational medicine for in-silico study of hematological phenomenon and disorders. For example, analysis of hemoglobin defects by applied usage of computational tools for nanohematology [45 - 48]. To help the reader better understand the concept, an example on the approach to determine the structural defect of hemoglobin Siam disorder will be presented. A computer-based study for amino acid sequence comparison and protein structure modeling is performed. The database ExPASY [49] was used for searching for the amino acid sequence of normal human alpha globin chain. Then the mutation alpha 15 Gly-->Arg was experimentally performed. Concerning secondary structure modeling, the author performed protein secondary structure predictions of alpha globins in both normal and hemoglobin Siam disorder from its primary sequence using NNPREDICT server [50]. The calculated secondary structures were presented and compared. The calculated tertiary structures were also presented and compared. All programs used in this study are standard programs used in bioinformatic study. Using NNPREDICT server, the calculation for secondary structure of alpha globin chains of normal and Hb Siam disorder was performed. Considering the derived secondary structures, there are 73 helix and 4 strands in normal alpha globin chain and there are 72 helix and 4 strands in alpha globin chain of Hb Siam. The difference of alpha globin chains between normal and Hb Siam disorder is presented in Figure 1.

A. alpha globin chain in normal

--------H--HHHHHHH---H-HHHHHHHHHHHH-------------------HEH---
-HHHHHHHHHHHH------HHHHHHHHHHHH---------HHHHHHHHHHHHHHH-------H
HHHHHHHHHHHHHHHEEE-----

B. alpha globin chain in hemoglobin Siam

--------H—HHH-HHH---H-HHHHHHHHHHHH-------------------HEH---
-HHHHHHHHHHHH------HHHHHHHHHHHH---------HHHHHHHHHHHHHHH-------H
HHHHHHHHHHHHHHHEEE-----

Figure 1. Calculated secondary structures of alpha globin chains of normal and Hb Siam disorder (Secondary structure prediction: *H = helix, E = strand, - = no prediction).*

In addition to simple prediction, interaction between nanomolecules can also be simulated. For example, Wiwanitkit performed a prediction on the binding affinity on the NS1 immunoglobulin (IgG and CD61-IgG) complexes using a computation medicine way [51]. According to this study, the significant discrimination of modeled Ig complexes could be detected [51]. The finding, which was expected for the late recovery phase of dengue, implied the difficulty in mimicking immune complex formation [51].

REFERENCES

[1] Waenlor W. Analysis of hemoglobin disorders by applied usage of computational tools for nanohematology. *Nanomedicine.* 2005 Sep;1(3):219.

[2] Freitas RA Jr. What is nanomedicine? *Nanomedicine.* 2005 Mar;1(1):2-9.

[3] Leary SP, Liu CY, Apuzzo ML. Toward the emergence of nanoneurosurgery: part II--nanomedicine: diagnostics and imaging at the nanoscale level. *Neurosurgery.* 2006 May;58(5):805-23.

[4] Harvey MA. Point-of-care laboratory testing in critical care. *Am. J. Crit. Care.* 1999 Mar;8(2):72-83.

[5] Muller MM, Hackl W, Griesmacher A. Point-of-care-testing--the intensive care laboratory. *Anaesthesist.* 1999 Jan;48(1):3-8.

[6] Giuliano KK, Grant ME. Blood analysis at the point of care: issues in applied usage for use in critically ill patients. *AACN Clin. Issues.* 2002 May;13(2):204-20.

[7] Halpern NA, Brentjens T. Point of care testing informatics. The critical care-hospital interface. *Crit. Care Clin.* 1999 Jul;15(3):577-91, vi-vii.

[8] Schallom L. Point of care testing in critical care. *Crit. Care Nurs. Clin. North Am.* 1999 Mar;11(1):99-106.

[9] Dirks JL. Diagnostic blood analysis using point-of-care technology. AACN Clin Issues. 1996 May;7(2):249-59.

[10] Geyer SJ. Implementing an effective point-of-care testing program *Crit. Care Nurs. Q.* 2001 May;24(1):7-14.

[11] Toner M, Irrimia D. BLOOD-ON-A-CHIP. *Ann. Rev. Biomed. Eng.* 2005 Aug ; 7: 77-103.

[12] Tai YC. Blood count on a chip. Proceedings of the 2nd IEEE International Conference on Nano/Micro Engineered and Molecular Systems, IEEE NEMS 2007.

[13] Kang Y, Li D, Kalams SA, Eid JE. DC-Dielectrophoretic dividing of biological cells by size. *Biomed. Biodevic.* 2007Sep 27

[14] Bardell RL, Weigl BH, Kesler N, Schulte T, Hayenga J, Battrell F. Microfluidic disposables for cellular and chemical detection - CFD model results and fluidic verification experiments Proceedings of SPIE - *The International Society for Optical Engineering* 4265, 1-13.

[15] Wiwanitkit V, Seereemaspun A, Rojanathanes R. Increasing the agglutination reaction in slide test for weak B blood group by gold nanoparticle solution: The first world report. *J. Immunol. Methods.* 2007 Oct 4;

[16] Wiwanitkit V, Soogarun S, Suwansaksri J. A correlative study on red blood cell parameters and urine trans, trans-muconic acid in subjects with occupational benzene exposure. *Toxicol. Pathol.* 2007; 35(2):268–269

[17] Soler MA, Bao SN, Alcantara GB, Tiburcio VH, Paludo GR, Santana JF, Guedes MH, Lima EC, Lacava ZG, Morais PC. Interaction of erythrocytes with magnetic nanoparticles. *J. Nanosci. Nanotechnol.* 2007; 7(3):1069–1071

[18] Rothen-Rutishauser BM, Schurch S, Haenni B, Kapp N, Gehr P. Interaction of fine particles and nanoparticles with red blood cells visualized with advanced microscopic ways. *Environ. Sci. Technol.* 2006; 40(14):4353–4359.

[19] Thrall L. Are red blood cells defenseless against smaller nanoparticles? *Environ. Sci. Technol.* 2006; 40(14):4327–4328.

[20] Wiwanitkit V. Visualization of gold nanoparticle on the microscopic picture of red blood cell: implication for probable risk of nanoparticle exposure. *Stoch. Env. Res. Risk Asses.* 24 Aug

[21] Siiman O, Gordon K, Burshteyn A, Maples JA, Whitesell JK. Immunophenotyping using gold or silver nanoparticle-polystyrene bead conjugates with multiple light scatter. *Cytometry.* 2000;41:298-307.

[22] Bene L, Szentesi G, Matyus L, Gaspar R, Damjanovich S. Nanoparticle energy transfer on the cell surface. *J. Mol. Recognit.* 2005;18:236-53.

[23] Pittet MJ, Swirski FK, Reynolds F, Josephson L, Weissleder R. Tagging of immune cells for in vivo imaging using magnetofluorescent nanoparticles. *Nat. Protoc.* 2006;1:73-9.

[24] Radomski A, Jurasz P, Alonso-Escolano D, Drews M, Morandi M, Malinski T, Radomski MW. Nanoparticle-induced platelet aggregation and vascular thrombosis. *Br. J. Pharmacol.* 2005 Nov;146(6):882-93.

[25] Haruta M. When gold is not noble: catalysis by nanoparticles. *Chem. Rec.* 2003;3(2):75-87.

[26] Cejas MA, Chen C, Kinney WA, Maryanoff BE. Nanoparticles That Display Short Collagen-Related Peptides. Potent Stimulation of Human Platelet Aggregation by Triple Helical Motifs. *Bioconjug. Chem.* 2007 Jun 21; [Epub ahead of print]

[27] Okamura Y, Handa M, Suzuki H, Ikeda Y, Takeoka S. New method of platelet substitutes for enhancing platelet aggregation at high shear rates: cooperative effects of a mixed system of fibrinogen gamma-chain dodecapeptide- or glycoprotein Ibalpha-conjugated latex beads under flow conditions. *J. Artif. Organs.* 2006;9(4):251-8.

[28] Sahli H, Tapon-Bretaudiere J, Fischer AM, Sternberg C, Spenlehauer G, Verrecchia T, Labarre D. Interactions of poly(lactic acid) and poly(lactic acid-co-ethylene oxide) nanoparticles with the plasma factors of the coagulation system. *Biomaterials.* 1997 Feb;18(4):281-8.

[29] Chen H, Titushkin I, Stroscio M, Cho M. Altered membrane dynamics of quantum dot-conjugated integrins during osteogenic differentiation of human bone marrow derived progenitor cells. *Biophys J.* 2007 Feb 15;92(4):1399-408.

[30] Garon EB, Marcu L, Luong Q, Tcherniantchouk O, Crooks GM, Koeffler HP. Quantum dot tagging and tracking of human leukemic, bone marrow and cord blood cells. *Leuk. Res.* 2007 May;31(5):643-51.

[31] Horak D, Babic M, Jendelova P, Herynek V, Trchova M, Pientka Z, Pollert E, Hajek M, Sykova E. D-mannose-modified iron oxide nanoparticles for stem cell tagging. *Bioconjug Chem.*. 2007 May-Jun;18(3):635-44.

[32] Zheng J, Ghazani AA, Song Q, Mardyani S, Chan WC, Wang C. Cellular imaging and surface marker tagging of hematopoietic cells using quantum dot bioconjugates. *Lab. Hematol.* 2006;12(2):94-8.

[33] Jordan A. Nanotechnology and consequences for surgical oncology. Kongressbd Dtsch *Ges. Chir. Kongr.* 2002;119:821-8

[34] Portney NG, Ozkan M. Nano-oncology: medicine delivery, imaging, and sensing. *Anal. Bioanal. Chem.* 2006 Feb;384(3):620-30.

[35] Maysinger D. Nanoparticles and cells: good companions and doomed partnerships. *Org. Biomol. Chem.* 2007 Aug 7;5(15):2335-42.

[36] Choi AO, Cho SJ, Desbarats J, Lovric J, Maysinger D. Quantum dot-induced cell death involves Fas upregulation and lipid peroxidation in human neuroblastoma cells. *J. Nanobiotechnology.* 2007 Feb 12;5:1.

[37] Schatzlein AG, Rutherford C, Corrihons F, Moore BD. Phage derived peptides for targeting of doxorubicin conjugates to solid tumours. *J. Control. Release.* 2001 Jul 6;74(1-3):357-62.

[38] He H, Conrad CA, Nilsson CL, Ji Y, Schaub TM, Marshall AG, Emmett MR. Method for Lipidomic Analysis: p53 Expression Modulation of Sulfatide, Ganglioside, and Phospholipid Composition of U87 MG Glioblastoma Cells. *Anal. Chem.* 2007 Oct 12;

[39] Li G, Huang J, Li Y, Zhang R, Deng B, Zhang J, Aoki H. In vitro study on influence of a discrete nano-hydroxyapatite on leukemia P388 cell property. *Biomed. Mater. Eng.* 2007;17(5):321-7.

[40] Luo LY, Zhang TL, Wang K. Differentiation of HL-60 cells induced by realgar nano-particles. *Zhongguo Zhong Yao Za Zhi.* 2006 Aug;31(16):1343-6.

[41] Lapotko DO, Lukianova E, Oraevsky AA. Selective laser nano-thermolysis of human leukemia cells with microbubbles generated around clusters of gold nanoparticles. *Lasers Surg .Med.* 2006 Jul;38(6):631-42.

[42] Xia Q, Wang HX, Wang J, Liu BY, Hu MR, Zhang XM, Shen BF. Nano-ESI-MS/MS identification on differentiation-associated proteins in M1 mouse myeloid leukemia cells induced by IL-6. *Zhongguo Yi Xue Ke Xue Yuan Xue Bao.* 2004 Oct;26(5):483-7.

[43] Rezaul K, Wu L, Mayya V, Hwang SI, Han D. A systematic characterization of mitochondrial proteome from human T leukemia cells. *Mol. Cell Proteomics.* 2005 Feb;4(2):169-81.

[44] Ning N, Peng ZF, Yuan L, Gou BD, Zhang TL, Wang K. Realgar nano-particles induce apoptosis and necrosis in leukemia cell lines K562 and HL-60. *Zhongguo Zhong Yao Za Zhi*. 2005 Jan;30(2):136-40.

[45] Wiwanitkit V. Modeling for tertiary structure of globin chain in Hemoglobin Suan-Dok disorder. *Hematology*. 2005 Apr;10(2):163-5.

[46] Wiwanitkit V. Secondary and tertiary structure aberration of alpha globin chain in haemoglobin Q-India disorder. *Indian J. Pathol. Microbiol.* 2006 Oct;49(4):491-4.

[47] Wiwanitkit V. Structural analysis on the abnormal elongated hemoglobin "hemoglobin Geneva". *Nanomedicine*. 2005 Sep;1(3):216-8.

[48] Wiwanitkit V. Tertiary structural analysis of the elongated part of an abnormal hemoglobin, hemoglobin Pakse. *Int. J. Nanomedicine*. 2006;1(1):105-7.

[49] Gasteiger E, Gattiker A, Hoogland C, Ivanyi I, Appel RD, Bairoch A. ExPASy: The proteomics server for in-depth protein knowledge and analysis. *Nucleic. Acids Res.* 2003;31:3784-8.

[50] Kneller DG, Cohen FE, Langridge R. Improvements in Protein Secondary Structure Prediction by an Amplified Neural Network. *J. Mol. Biol.* 1990; 214: 171-182.

[51] Wiwanitkit V. Weak binding affinity of immunoglobin G, an explanation for the immune mimicking theory in pathophysiologic findings in the recovery phase of dengue. *Nanomedicine*. 2005 Sep;1(3):239-40.

Chapter 10

NANOINFORMATICS

INTRODUCTION TO BIOINFORMATICS

Genetics began when Mendel proved his laws of heredity with varieties of peas and flowers in 1865 [1]. This is the basis that led to the sequencing of the first complete genome (*Haemophilus influenzae*) published in 1995 [2]. We have since moved on to technologies allowing the sequencing, recombination and cloning of DNA. The unveiling of the Human Genome Project (HGP) by the United States Department of Energy (DoE) and the National Institutes of Health is the big jumping-off point [3 - 4]. Goals were to identify all chemical base pairs and all genes that make up the 23 chromosome pairs found in human DNA [4]. The main aims also included producing new laboratory and computing technologies to make all this probable and disseminating genome information [4]. This led to the post-genome era, the bioinformatic era. Bioinformatics has been split into various subjects: many omics. Bioinformaticists act to bridge the gap between the data stored and its biological significance. The first process for bioinformatics is data mining [5 - 6]. Passive ways have to know and understand how to use those methods and how to interpret the result. Various methods in statistics, namely, Min, Max, SD, Regression and correlation are used. While active ways especially for artificial intelligence needs no such complicated long protocol. There is no need for the user to set up any criteria or hypothesis whether which data should be in the same group or correlate to those groups. This new way can lead the user to find an unpredicted outcome.

Of many omics sciences, one of the most important is genomics. Genomics is also the first omics science. Genomics [7 – 9] refers to the analysis of all of the genes and transcripts included within the genome. For comparative genomics [10 - 11], complete genomic sequences are now available for many organisms/bacteria/viruses/organelles including the following 'model' organisms (Table 1). In addition to comparative genomics, functional genomics [12 - 13] refers to the development and applied usage of global (genome-wide or system-wide) experimental approaches to assess gene function by making use of the information and reagents bringd by structural genomics. It is characterized by high-throughput or large-scale experimental methodologies combined with statistical or computational analysis of the results. Computational biology strives to extract the maximal probable information from known sequences, by classifying them according to their homologous relationships, predicting their biochemical activity, cellular function, 3-dimensional structures and evolutionary starting point. Phylogenetic profile [14] is the good method of functional genomics. The phylogenetic profile describes the pattern of existence or absence of a particular protein, across a set of organisms. This method is useful to study the relation among the germs from different settings and can track back to its starting point. In addition, complex profiles can lead to the more complicated network.

Table 1. What to compare in comparative genomics

Items	Details
Gene site	Site
Gene structure	Exon number
	Exon lengths
	Intron lengths
	Sequence similarity
Gene characteristics	Splice sites
	Codon usage
	Conserved synteny

Database in Nanoscience

The summation of biomedical information grows day by day. At present, collection and systematization of the medical information are developed from collections of hardcopies in a library to the collections of electronic data in computers. Computer applied usages in medicine start from the point of view of the levels of information processing as suggested by van Bemmel and go into the present research directions. Present applied usages cover the subjects of communication, databases, complex calculations (including signal analysis and image analysis ways), pattern recognition and expert systems [15]. Hasman said that the role the computer system played in relation to the physician could not replace the physician because computer systems used programs that were of a scientific nature. However, the applied usage of a computer, especially database technology, brings a lot of help to physicians and medical scientists [15]. It is accepted that the use of computer technology in medicine is no longer the domain of only a few "gadobtain" happy high-tech aficionados but the rapid pace of medical progress [16]. Presently, there are a number of generated databases. Some are accessible by the public, others are private. The usage of a public database can enhance the quality and effectiveness of patient care and the health of the public. Renschler said that publications could be retrieved and downloaded anywhere and anytime with the introduction of electronic publishing and full-text databases becoming available [17]. In addition, study groups for practice-based learning can prepare themselves for discussions of their problems or of simulated cases systematically bringd by central organizations with experts using information technology [17]. Renschler also mentioned that a pilot study revealed great interest in the applied usage of information technology; most of the responding colleagues revealed interest in occasional or regular use of medical or non-medical full text database, preferably using their own computers [17]. In nanoscience, there are also many new developed databases. Here, the author will focus on some important public databases.

1. Nanobase

Nanobase is a new database that acts as a search engine for the nanomaterial area. The search covers data on fullerene, carbon nanotube, nanoparticles, quantum dot, nanowire and nanofiber. This tool is available online at www.nanowerk.com. Nanobase claims it is the world's best source for finding nanomaterials from worldwide nanotechnology suppliers. The database consist ofs at least 1,726 nanoparticles from 122 suppliers.

2. Nanotechnology Database

This site is designed to present up-to-date sources of information on nanotechnology in these areas: major research centers, funding agencies, major reports and books. The summaries which appear were taken directly from each of their respective sites. This tool is available online at www.wtec.org.

3. NSTI Nanotechnology Database

NSTI Technology Database collects information on conferences, publications and network of advisors to continuously assess the state-of-the-art in emerging technologies and in industrial experiment sector needs. This high-quality information is classified and archived in a powerful relational database according to technical capabilities, intellectual property, and business interests and encompasses at least 6,000 direct contacts (detailed), 1,400 corporations (detailed), and 43,000 opt-in email contacts. This tool is available online at http://www.nsti.org.

4. Nanoethicsbank

The NanoEthicsBank is a database that is being constructed with the aim of being a resource for researchers, scholars, students, and the general public who are interested in the social and ethical implications of nanotechnology. Collected items in the database include normative documents, such as guidelines for safety in the workplace, and descriptive materials and studies of the media coverage of nanotechnology. This tool is available online at http://hum.iit.edu/NanoEthicsBank.

Genomics Study in Nanoscience

Functional genomics as a standard tool helped to identify the function of many of the genes that have encoded important transporters and metabolic enzymes over the past few years [18]. Advanced expression systems and analysis technologies made it probable to study the biochemical properties of the corresponding proteins in great detail [18]. The present nanotechnology allows us to develop new methods to separate many biomolecules, including DNA, RNA, protein, peptide, polysaccharide, and glycoprotein towards the applied usage in genomics, transcriptomics, proteomics, and glycoproteomics [19]. Examples of interesting reports on genomics study in nanoscience are presented in Table 2.

Table 2. Examples of interesting reports on genomics study in nanoscience

Authors	Details
Boccazzi et al. [20]	Boccazzi et al. compared global gene expression using 500 ng of total RNA from *Escherichia coli* cultures developed in rich or defined minimal media in a miniaturized 50-microl bioreactor [20]. According to this work, the results from microscale cultures using just 500 ng of total RNA indicated that high-throughput integration of growth physiology and genomics would be probable with new biochemical platforms and improved detection technologies [20].
Tian et al. [21]	Tian et al. described a new microchip-based technology for multiplex gene synthesis [21]. Tian et al. concluded that this technology enabled scientists to synthesize all 21 genes that encode the proteins of the *Escherichia coli* 30S ribosomal subunit, and to optimize their translation efficiency in vitro through alteration of codon bias [21].
Karlin-Neumann et al. [22]	Karlin-Neumann et al. found that quantum dots bringd a full alternative to organic dyes for achieving multiplexed and accurate, four-color genotyping and have the additional usefulness of being excitable by a single wavelength of light despite their distinct emission wavelengths [22].

PROTEOMICS STUDY IN NANOSCIENCE

Proteomics study in nanoscience is generally mentioned more than genomics study in nanoscience. This might be because nanoscience involves the small particles which are more common to the structure than genomics codons. Based on the comparatives and functional proteomics, there is much research in the field of nanoscience.

A. Research on the Applied Proteomics Ways

There is much research on applied proteomics ways for diagnosis. In 2006, Yu et al. presented the first example of the analysis of acidic and basic proteins by nanoparticle-filled capillary electrophoresis [23]. According to this work, compared to the didodecyldimethylammonium bromide-coated capillary, the DDAB-capped gold nanoparticles as pseudostationary phase were found to form more stable coating on the capillary wall, thus leading to greater dividing efficiency and high reproducibility [23]. Yu et al. proposed that with simplicity, high resolving power, and high reproducibility, the proposed method could have great potential for proteomics applied usages and clinical diagnosis [23]. Becker et al. developed a robust, fast, standardized, and easy-to-use way for the solubilization of non-degraded, full length, and immunoreactive proteins from formalin-fixed tissues for western blot and protein microarray analysis [24]. According to the experiment of Becker et al., they succeeded in isolating non-degraded, soluble, and immunoreactive proteins from routinely processed formalin-fixed tissues and were able to detect membrane, cytoplasmic and nuclear proteins at the expected molecular weight [24]. Becker et al noted that a new clinically orientated multiplexed protein measurement system might be generally applied used to determine the relative abundances of known disease-related proteins in small summations of routinely processed formalin-fixed tissue samples for research and diagnosis [24]. Examples of other interesting reports on proteomics study in nanoscience are presented in Table 3.

Table 3. Reports on proteomics study on the applied proteomics ways in nanoscience

Authors	Details
Nagele et al. [25]	Nagele et al. developed a new method for 2 dimensional online nano-liquid chromatography/mass spectrometry that replaced the injected salt step gradient with an optimized semicontinuous pumped salt gradient to improve the characteristic of strong cation exchange chromatography [25]. According to this work the semicontinuous pump salt method compared designedably with the commonly used injection method and also with an offline 2 dimensional - liquid chromatography method [25].
Winnik [26]	In this work, a new empirical strong cation exchange peptide elution score was developed, which accounted for both the number of basic and acidic residues and, in part, their site within a sequence of a peptide [26].
Vollmer et al. [27]	Using a nano-high-characteristic liquid chromatography/mass spectrometry system Vollmer et al. constructed comprehensive peptide maps for *Escherichia coli* developed with either lactose or glucose in minimal medium [27].
Nagele et al. [28]	Nagele et al. demonstrated the development of a method for the analysis of complex proteome samples by two-dimensional nano-liquid chromatography-mass spectrometry [28]. This approach includes strong cation-exchange, sample enrichment, reversed-phase chromatography and nanospray ion trap mass spectroscopy with data dependent tandem mass spectrometry spectra acquisition, and subsequent database search [28].
Ihling et al. [29]	Ihling et al. reported the applied usage of a nano- high-characteristic liquid chromatography system coupled to a Fourier transform ion-cyclotron resonance mass spectrometer equipped with nanoelectrospray source for proteome analysis [29].
Williams and Tomer [30]	Williams and Tomer reported the incorporation of on-line sample clean-up and analyte concentration with a high-throughput, chip-based, robotic nano-ESI-mass spectrometry platform for proteomics studies [30].
Michel et al. [31]	Michel et al. presented a comparative proteomic analysis of human maternal plasma and amniotic fluid samples from the same patient at term of pregnancy in order to find specific amniotic fluid proteins as markers of premature rupture of membranes, a complication frequently observed during pregnancy [31]. As a further validation of the method, plasma proteins fractionated by off-gel electrophoresis and analyzed by nano-liquid chromatography-mass spectrometry/mass spectrometry had also been compared to the Swiss 2-dimensional PAGE reference map by reconstructing a map that matches 2-dimensional gel and off-gel electrophoresis experimental data [31]. Michel et al. revealed that most reference proteins could be identified in both data sets, and that isoform shifts in pI are well conserved in the off-gel electrophoresis data sets [31].

B. Research on the Structures of Nano-Objects

As mentioned in the first chapter of this book, nano-objects are very small objects. The sizes of small protein and antibody are also at the nanolevel. Therefore, the applied usage of structural proteomics for answering questions on the structures of small proteins is probable. Interesting studies in this area are quoted in Table 4.

Table 4. Examples of interesting reports on proteomics study in nanoscience on the structures of nano-objects

Authors	Details
Wiwanitkit [32]	In this study the amino acid sequence of human alpha globin was extracted using ExPASY and compared with that obtained from the Hemoglobin (Hb) Suan-Dok disorder [32]. The derived sequences, alpha globin chains in both the normal and Hb Suan-Dok disorder, were used for further investigation of the tertiary structures [32]. Modeling these proteins for the tertiary structure was performed using the CPHmodels 2.0 Server [32]. For comparison the tertiary structure of human alpha globin chains in normal and hemoglobin Suan-Dok were calculated and presented [32]. Based on this information, there was no significant difference between the predicted alpha globin tertiary structures of normal hemoglobin and Hb Suan-Dok [32].
Wiwanitkit [33]	In this study, amino acid sequence of human alpha globin was searched using ExPASY and used for further mutation to Hb Q-India disorder [33]. The derived sequences, alpha globin chains in both normal and Hb Q-India disorder, were used for further investigation for secondary and tertiary structures [33]. Modeling of these proteins for secondary and tertiary structures was done using the NNPREDICT server and CPHmodels 2.0 Server, respectively [33]. In this study, the secondary and tertiary structures of human alpha globin chains of normal and HbQ-India disorder are calculated and presented [33]. Based on this information, the main difference between the predicted alpha globin secondary structures of normal and Hb Q-India is an extra helix in the Hb Q-India [33]. The predicted tertiary structure also supports this finding [33]. The results from this study can be good data for further study on Hb Q-India disorder, which can bring further understanding on this hemoglobinopathy [33].
Wiwanitkit [34]	In this report, a bioinformatic analysis was performed to study the secondary and tertiary structures of those abnormal amino acid sequences of Hb Geneva [34]. A computer-based study for protein structure modeling was performed [34]. According to this study, the secondary structure analysis of the Hb Geneva revealed many defects in helix and strand of the Hb Geneva compared with normal beta-globin chains [34]. On the basis of this information, the main alteration in the Hb Geneva might be due to these aberrations [34]. With regard to the tertiary structure, the deterioration of folds, accompanied by the aberration in secondary structure of globin in Hb Geneva can be identified [34].
Wiwanitkit [35]	In this work, a computer-based protein structure modeling was used in a bioinformatics analysis of the tertiary structure of these elongated amino acid sequences of HbPakse [35]. The elongated part of Hb Pakse revealed additional helices, which may cause the main alteration in Hb Pakse [35]. Abnormalities in the fold structure of globin in Hb Pakse were identified, and helices additional to the normal alpha globin chains were shown in the elongated part of Hb Pakse [35].

Authors	Details
Wiwanitkit [36]	A functional analysis was performed on 4 important beta hemoglobinopathies (Hb C, D, E, and S) using PolyPhen, a new bioinformatic tool [36]. The mutations Hb C (beta 6, Glu --> Lys), Hb D (beta 121, Glu --> Gln), Hb E (beta 26, Glu --> Lys), and Hb S (beta 6, Glu --> Val) were selected for further study [36]. According to the in silico mutation study, the functional change in the studied hemoglobinopathies was variable [36]. The position-specific independent counts (PSIC) difference score ranged from 1.362 (Hb D) to 2.986 (Hb S) [36]. Regarding the degree of damage, all had probable damage [36]. This analysis demonstrated that the functional aberration in the hemoglobinopathy was based on complex germesis [36]. Identifying only the structural aberration in a hemoglobinopathy is not sufficient; additional functional analysis is recommended [36].
Wiwanitkit [37]	Wiwanitkit performed a bioinformatic analysis to determine positions that tend to comply with peptide motifs in the amino acid sequence of hepatitis C virus (HCV) protein kinase -eIF2-alpha phosphorylation homology domain (PePHD) [37]. To identify weak linkage in HCV PePHD, a new bioinformatic tool, GlobPlot, was used [37]. Positions 2-7, 29-39, 53-57, 90-98, 123-133, 202-227, 324-355, 439-448 were identified as positions resistant to mutation [37]. Some are already known and others are newly discovered [37]. Based on this study, weak linkages in HCV PePHD could be identified and can be good information for expectation of probable new mutations that can lead to failure of HCV therapy [37].

Expression Study in Nanoscience

The study of the finalized expression of gene and protein is the wide focus in present translational research. Advances in nanoscience are having a significant impact on the field of biodiagnostics, where a number of nanoparticle-based assays have been introduced for biomolecular detection, with DNA- or protein-functionalized gold nanoparticles used as the target-specific probes [38]. The applied usage of nanobiotechnology in microarray systems is the best example of expression study in nanoscience. At present, expression analysis is routinely based on DNA microarrays due to highly parallel readout and high throughput. The marriage of microarray technologies with the emerging field of microfluidics brings a number of usefulness such as, reduction in reagent cost, reductions in hybridization assay times, high-throughput sample processing, and integration and automation capabilities of the front-end sample processing steps [39]. In addition, photoimmobilization method was developed for the preparation of microarray biochips [40]. It is accepted that the photoimmobilization method makes it probable to easily covalently immobilize various types of organic molecules and cells on a chip [40]. In addition, by using hydrophilic polymers as matrixes, it is probable to reduce nonspecific interactions with biological components [40]. Interesting studies in this area are quoted in Table 5.

Table 5. Examples of interesting reports on expression study in nanoscience

Authors	Details
Sunkara et al. [41]	Sunkara et al. reported the utility of the nano-scale controlled surface in conjunction with a multiply tagged protein [41]. According to this work, applied usage of streptavidin-fluorophore conjugates in combination with the highly controlled surface that suppresses non-specific binding of DNA allows highly sensitive detection of DNA while maintaining superior SNP discrimination efficiency comparable to our earlier results [41]. In addition, the sensitivity of DNA microarray on the mesospaced surface was two orders of magnitude higher than that of the generic surface when a streptavidin-fluorophore conjugate was employed, and the detection limit on the former surface was found to be 50 fM of 15-mer target DNA [41].
Terry et al. [42]	Terry et al. reported how the use of a standard integrated circuit fabrication process could improve the potential of silicon nitride layers as substrates for microarray technology [42]. Terry et al. said that the results for the plasma-amplified chemical vapor deposition silicon nitride were particularly significant for future devices that hope to integrate microelectronic control and analysis circuitry, due to the film's use as a final passivating layer [42].
Zhang et al. [43]	Zhang et al. proposed a new computational method to geometric measure of biological particles [43]. In this paper, the new method was applied to DNA microarray spot size estimation and the proposed method was found to be a computational efficient and also brings confidence probability on the measurement [43].
Dahl et al. [44]	Dahl et al. revealed the quantification ability and reliability of qPCR in 200 nl with the microPCR chip down to 5 starting target molecules using TaqMan chemistry [44]. Dahl et al. noted that the high sensitivity and quantification capability of the microPCR chip platform developed in this work made it a useful technology for performing high-throughput qPCR-based analysis in the nanoliter volume range [44].

SIMULATION ANALYSIS BY NANOINFORMATICS

Simulation analysis is a useful way to perform in silico experiments. Simulation analysis can be effectively performed by the present advanced bioinformatics ways. At present, there are also many new specific nanoinformatics tools for simulation analysis. The nanoHub is the specific database collecting basic tool for the simulation purpose. It is accepted that NanoHUB is a virtual computing center created to support nanotechnology research and is the excellent gateway for nanoscience. The three main simulation areas according to nanoHub are a) simulation for nanoelectronics, b) simulation for NEMS and Nanofluidics and c) simulation for nano-bio devices. The way is rather new therefore there are only a few reports in this area. Wiwanitkit detected the oxidation flux change in diabetic nephropathy [45]. In this work, a simulation test to determine the oxidation flux change based on nano medicine way was used [45]. An increase in oxidation flux in the diabetic nephropathy could be derived [45]. Therefore, this work can support the finding that the oxidation flux change plays an important role in the germesis of diabetic nephropathy [45]. To help the reader to better understand the way, an example on simulation analysis study by nanoinformatics is hereby presented. The author hereby performed a simulation study to access the oxidation flux change on spermatozoa membrane in hypoxia. The simulation test to determine the oxidation flux change based on nanomedicine way is used. The way namely "PROCESS: OXIDATION FLUX" was used. Briefly, this simulation way integrates both the classic Deal-Grove's model and Massoud's model, which both describe the oxidation growth process [46]. Specifically, this way investigates the effect of different parameters and conditions on the oxidation process by looking into the oxidation flux [46]. It gives users the freedom to modify critical parameters and conditions in the process, such as oxidant condition, time, initial oxide thickness, temperature, pressure, crystal orientation, too as an opportunity to choose between the Deal-Grove's or Massoud's model, or a combination of both [46]. The primary condition in this study is wet condition, temperature = 37 degree Celcius, run time = 60 minutes, and thickness = 3000 nm [47] according to the normal physiological condition. According to the simulation test, the oxidant concentration for normal equal to 3e+19/cm^3 when varying the oxygen pressure from 0 to 100 mmHg. Of interest, change of flux can be detected. The main germesis should be the direct injury of membrane structure of spermatozoa by free radicals which can lead to sperm defect. Therefore, this work can support the finding that the oxidation flux change corresponding to oxygen pressure change in spermatozoa does not exist. However, the flux change can be seen if the membrane thickness of spermatozoa is varied. Thin membrane spermatozoa is more prone to oxidative stress than thick membrane ones. The defect in the enzymatic system within the spermatozoa should be a better explanation for vulnerability of spermatozoa to oxidative stress. The use of enzymatic modification way by antioxidants can be a useful alternative in the management of male infertility [48 - 49].

REFERENCES

[1] Trgovcevic Z. 50 years of molecular biology. *Lijec Vjesn.* 1994 Nov-Dec;116(11-12):315-8.

[2] Fleischmann RD, Adams MD, White O, Clayton RA, Kirkness EF, Kerlavage AR, Bult CJ, Tomb JF, Dougherty BA, Merrick JM, et al Whole-genome random sequencing and assembly of Haemophilus influenzae Rd. *Science.* 1995 Jul 28;269(5223):496-512.

[3] Lewin R. National Academy looks at human genome project, sees progress. *Science.* 1987 Feb 13;235(4790):747-8.

[4] Johson RS. The human genome project: what impact on basic research? *FASEB J.* 1987 Dec;1(6):502-5.

[5] Yang ZR, Hamer R. Bio-basis function neural networks in protein data mining. *Curr. Pharm. Des.* 2007;13(14):1403-13.

[6] Ahrens CH, Wagner U, Rehrauer HK, Turker C, Schlapbach R. Present challenges and approaches for the synergistic use of systems biology data in the scientific community. *EXS.* 2007;97:277-307.

[7] Jordan BR. 'Genomics': buzzword or reality? *J. Biomed. Sci.* 1999 May-Jun;6(3):145-50

[8] Koning DJ, Haley CS. Genetical genomics in humans and model organisms. *Potentials Genet.* 2005 Jul;21(7):377-81.

[9] Claverie JM, Abergel C, Audic S, Ogata H Recent advances in computational genomics. *Pharmacogenomics.* 2001 Nov;2(4):361-72.

[10] O'Brien SJ, Wienberg J, Lyons LA. Comparative genomics: lessons from cats. *Potentials Genet.* 1997 Oct;13(10):393-9.

[11] O'Brien SJ, Menotti-Raymond M, Murphy WJ, Nash WG, Wienberg J, Stanyon R, Copeland NG, Jenkins NA, Womack JE, Marshall Graves JA. The promise of comparative genomics in mammals. *Science.* 1999 Oct 15;286(5439):458-62, 479-81.

[12] Evans MJ, Carlton MB, Russ AP. Gene trapping and functional genomics. *Potentials Genet.* 1997 Sep;13(9):370-4.

[13] Rastan S, Beeley LJ. Functional genomics: going forwards from the databases. *Curr. Opin. Genet Dev.* 1997 Dec;7(6):777-83.

[14] Pollock DD. Genomic biodiversity, phylogenetics and coevolution in proteins. *Appl. Bioinformatics.* 2002;1(2):81-92.

[15] Hasman A. Medical applied usages of computers: an overview. *Int. J. Biomed. Comput.* 1987; 20(4):239-51.

[16] American Medical Association. Medical informatics: an emerging medical discipline. Council on Scientific Affairs and Council on Long Range Planning and Development of the American Medical Association. *J. Med .Syst.* 1990; 14(4):161-79.

[17] Renschler HE. Rational continuing medical education. *Schweiz Rundsch Med. Prax.* 1991; 80(19):515-23.

[18] Deuschle K, Fehr M, Hilpert M, Lager I, Lalonde S, Looger LL, Okumoto S, Persson J, Schmidt A, Frommer WB. Genetically encoded sensors for metabolites. *Cytometry* A. 2005 Mar;64(1):3-9.

[19] Baba Y. Diagnosis of diseases by nanodevice. *Nippon Rinsho.* 2006 Feb;64(2):271-7.

[20] Boccazzi P, Zanzotto A, Szita N, Bhattacharya S, Jensen KF, Sinskey AJ. Gene expression analysis of Escherichia coli developed in miniaturized bioreactor platforms for high-throughput analysis of growth and genomic data. *Appl. Microbiol. Biotechnol.* 2005 Sep;68(4):518-32.

[21] Tian J, Gong H, Sheng N, Zhou X, Gulari E, Gao X, Church G. Accurate multiplex gene synthesis from programmable DNA microchips. *Nature.* 2004 Dec 23;432(7020):1050-4.

[22] Karlin-Neumann G, Sedova M, Falkowski M, Wang Z, Lin S, Jain M. Applied usage of quantum dots to multicolor microarray experiments: four-color genotyping. *Methods Mol. Biol.* 2007;374:239-52.

[23] Yu CJ, Su CL, Tseng WL. Dividing of acidic and basic proteins by nanoparticle-filled capillary electrophoresis. *Anal. Chem.* 2006 Dec 1;78(23):8004-10.

[24] Becker KF, Schott C, Hipp S, Metzger V, Porschewski P, Beck R, Nahrig J, Becker I, Hofler H. Quantitative protein analysis from formalin-fixed tissues: implications for translational clinical research and nanoscale molecular diagnosis. *J. Pathol.* 2007 Feb;211(3):370-8.

[25] Nagele E, Vollmer M, Horth P. Improved 2D nano-LC/MS for proteomics applied usages: a comparative analysis using yeast proteome. *J. Biomol. Tech.* 2004 Jun;15(2):134-43.

[26] Winnik WM. Continuous pH/salt gradient and peptide score for strong cation exchange chromatography in 2D-nano-LC/MS/MS peptide identification for proteomics. *Anal. Chem.* 2005 Aug 1;77(15):4991-8.

[27] Vollmer M, Nagele E, Horth P. Differential proteome analysis: two-dimensional nano-LC/MS of E. coli proteome developed on different carbon sources. *J. Biomol. Tech.* 2003 Jun;14(2):128-35.

[28] Nagele E, Vollmer M, Horth P. Two-dimensional nano-liquid chromatography-mass spectrometry system for applied usages in proteomics. *J. Chromatogr.* A. 2003 Aug 15;1009(1-2):197-205.

[29] Ihling C, Berger K, Hofliger MM, Fuhrer D, Beck-Sickinger AG, Sinz A. Nano-high-characteristic liquid chromatography in combination with nano-electrospray ionization Fourier transform ion-cyclotron resonance mass spectrometry for proteome analysis. *Rapid Commun. Mass Spectrom.* 2003;17(12):1240-6.

[30] Williams JG, Tomer KB. Disposable chromatography for a high-throughput nano-ESI/MS and nano-ESI/MS-MS platform. *J. Am. Soc. Mass Spectrom.* 2004 Sep;15(9):1333-40.

[31] Michel PE, Crettaz D, Morier P, Heller M, Gallot D, Tissot JD, Reymond F, Rossier JS. Proteome analysis of human plasma and amniotic fluid by Off-Gel isoelectric focusing followed by nano-LC-MS/MS. *Electrophoresis.* 2006 Mar;27(5-6):1169-81.

[32] Wiwanitkit V. Modeling for tertiary structure of globin chain in Hemoglobin Suan-Dok disorder. *Hematology.* 2005 Apr;10(2):163-5.

[33] Wiwanitkit V. Secondary and tertiary structure aberration of alpha globin chain in haemoglobin Q-India disorder. Indian J Pathol Microbiol. 2006 Oct;49(4):491-4.

[34] Wiwanitkit V. Structural analysis on the abnormal elongated hemoglobin "hemoglobin Geneva". *Nanomedicine.* 2005 Sep;1(3):216-8.

[35] Wiwanitkit V. Tertiary structural analysis of the elongated part of an abnormal hemoglobin, hemoglobin Pakse. *Int. J. Nanomedicine.* 2006;1(1):105-7.

[36] Wiwanitkit V. Analysis of functional aberration of some important beta hemoglobinopathies (hemoglobin C, D, E, and S) from nanostructures. *Nanomedicine.* 2005 Sep;1(3):213-5.

[37] Wiwanitkit V. Weak linkage in hepatitis C PePHD: identification of mutation prone point that can lead to failure of antiviral therapy for prevention of hepatocellular carcinoma. *Asian Pac. J. Cancer Prev.* 2007 Jan-Mar;8(1):139-40.

[38] Thaxton CS, Georganopoulou DG, Mirkin CA. Gold nanoparticle probes for the detection of nucleic acid targets. *Clin. Chim. Acta.* 2006 Jan;363(1-2):120-6.

[39] Situma C, Hashimoto M, Soper SA. Merging microfluidics with microarray-based bioassays. *Biomol. Eng.* 2006 Oct;23(5):213-31.

[40] Ito Y. Photoimmobilization for microarrays. *Biotechnol. Prog.* 2006 Jul-Aug;22(4):924-32.

[41] Sunkara V, Hong BJ, Park JW. Sensitivity enhancement of DNA microarray on nano-scale controlled surface by using a streptavidin-fluorophore conjugate. *Biosens Bioelectron.* 2007 Feb 15;22(7):1532-7.

[42] Terry JG, Campbell CJ, Ross AJ, Livingston AD, Buck AH, Dickinson P, Mountford CP, Evans SA, Mount AR, Beattie JS, Crain J, Ghazal P, Walton AJ. Improved silicon nitride surfaces for next-generation microarrays. *Langmuir.* 2006 Dec 19;22(26):11400-4.

[43] Zhang M, Mao K, Tao W, Tarn TJ. A computational method to geometric measure of biological particles and applied usage to DNA microarray spot size estimation. *Med. Biol. Eng. Comput.* 2006 Apr;44(4):275-9.

[44] Dahl A, Sultan M, Jung A, Schwartz R, Lange M, Steinwand M, Livak KJ, Lehrach H, Nyarsik L. Quantitative PCR based expression analysis on a nanoliter scale using polymer nano-well chips. *Biomed. Microdevices.* 2007 Jun;9(3):307-14.

[45] Wiwanitkit V. Oxidation flux change on glomerulus membrane in normal and diabetic nephropathy. *Ren. Fail.* 2007;29(5):549-51.

[46] Qiao W, McLennan M, Kennell R, Ebert DS, Klimeck G. Hub-based simulation and graphics hardware accelerated visualization for nanotechnology applied usages. *IEEE Trans Vis. Comput. Graph.* 2006;12:1061-8

[47] Nebel RL, Saacke RG. Technology and applied usages for encapsulated spermatozoa. *Biotechnol. Adv.* 1994;12:41-8.

[48] Gavella M, Lipovac V, Vucic M, Sverko V. In vitro inhibition of superoxide anion production and superoxide dismutase activity by zinc in human spermatozoa. *Int. J. Androl.* 1999;22:266-74.

[49] Storey BT. Biochemistry of the induction and prevention of lipoperoxidative damage in human spermatozoa. *Mol. Hum. Reprod.* 1997;3:203-13.

Chapter 11

NANOCOMPUTING

WHAT IS NANOCOMPUTING? [1 - 2]

The computer is a very useful scientific device. Similar to other scientific devices, as the years have gone by, computers have become smaller and smaller. The purpose of the small object is the ease in delivery. This concept can also be applied to the computer. One of the most important benefits of having smaller computers is convenience. Handheld compact disc players, MP3 players, and an iPod are basic commonly used small electronic devices. Combinations of the electronic devices into one piece such as a combination of the mobile telephone, television, camera and radio can be seen. The reduction in size of modern-day equipment is an important applied usage for the field of nanocomputing. One of the present goals for computer engineers is to make computers more qualitative and smaller.

Indeed, engineers realized the draft for how to do this a few decades ago but never could develop it due to resource limitation and limitation of size technology. At first, a computer had a very big size and required a lot of space. However, due to the advent of nanotechnology, it is easy to generate a very small nano-level piece, and this can reduce the size of the whole object. This can allow computer engineers to redesign computers at so small of a level that computers could be just as powerful as anyone had ever imagined. With reduced size, computers will be able to perform many complicated operations that will allow them to choose which calculation they should use. In the next step, computers might be able to make decisions instead of just listening to the user.

NANOTECHNOLOGY FOR NANOCOMPUTING

By definition, nanocomputing involves nano-scale computing. It involves the applied usage of the distinct properties of structures on the scale between 1 and 100 nm to solve complex mathematical or logical problems, or to accomplish any of the tasks that novel computers have the ability to do. Many advents in different nanotechnological fields help in the development of nanocomputing. Firstly, general desktop computing can be downsized based on nanotechnology, with the advent of nanotransistors. The long history of computer technology has seen a sequence of changes from one type of physical realisation to another from mechanical gears to relays to valves to transistors to integrated circuits and so on. Previously, the main obstacle to downsizing of the computer was the limitation in the size of transistors. Presently, advanced lithographic methods can create chips with features only a fraction of a micron or nano level wide. Nanotransistor is a kind of transistor that acts as an electronic signal switch or amplifier that has its size in nano-scale. There are many advents in nanotransistors at present. Some interesting reports on nanotransistors are hereby summarized in Table 1. Also, carbon nanotubes and silicon nanowires are available for use in computers to speed up connections between nanotransistors. There are also some other very interesting theoretical computing possibilities, the most important of which is quantum computing, which uses the quantum properties of single atoms to process information, which is more complex than the present ability of the classical computer. The details of some important ways that can be useful for nanocomputing will be further discussed.

Table 1. Some interesting reports on nanotransistors

Authors	Details
Hamann and Muller [3]	Hamann and Muller reported on absolute and approximate calculations of electron-energy-loss spectroscopy edge thresholds [3]. Hamann and Muller said that core level shifts could be predicted from ground-state eigenvalue differences only when comparing environments of similar electronegativity, and such was the case for the O K edge, but not the Si L edge at Si/SiO(2) interfaces in nanotransistors [3].
Appenzeller et al. [4]	Appenzeller et al. studied tunneling versus thermionic emission in one-dimensional semiconductors [4]. Appenzeller et al. reported on the role of contacts and the influence of Schottky barriers on the switching in nanotransistors [4]. They also discussed the factors that impacted tunneling probability with emphasis on the importance of the effective mass for transistor applied usages [4].
Jung et al. [5]	Vertically aligned carbon-nanotube (CNT) arrays were fabricated in the thin-film anodic aluminum oxide (AAO) templates on silicon wafers utilizing a niobium (Nb) thin film as the source electrode in this work [5]. Jung et al. reported that this structure could be a useful cornerstone in the fabrication of nanotransistors operating at room temperature [5].

A. Carbon Nanotube

Electronics obtained through the bottom-up approach of molecular-level control of material composition and structure may lead to devices and fabrication strategies not probable with top-down methods [6]. With the bottom-up approach, electromechanical and resistance-change memory devices based on carbon nanotubes can be performed. Since the discovery of the carbon nanotube by Iijima in 1991 [7], carbon nanotube have surged to the forefront as nanostructured material in nanoelectronic applied usages [8]. It is accepted that the carbon nanotube is useful for the development of a biosensor. The carbon nanotube can be divided into two types, single-walled and multiple-walled carbon nanotubes [9 - 12]. A common feature of the single-walled carbon-nanotube field-effect transistors fabricated to date has been the existence of a Schottky barrier [13]. These energy barriers severely limit transistor conductance and reduce the present delivery capability [13]. High-characteristic ballistic nanotube field-effect transistors with zero or slightly negative Schottky barriers were then developed by Javey et al [13]. More details on carbon nanotubes can be seen in another chapter in this book.

Table 2. Some interesting reports on carbon nanotubes for nanotransistors

Authors	Details
Gui et al. [14]	Gui et al. reported on the sensing mechanism of electrical detection of DNA hybridization for Au- and Cr-contacted field effect transistors based on single-walled carbon nanotube networks [14]. Gui et al. bringd comprehensive evidence that the sensing of DNA was dominated by the change in metal- single-walled carbon nanotube junctions rather than the channel conductance [14].
Ahn et al. [15]	Ahn et al. used scanning photopresent microscopy to investigate the properties of internal p-n junctions in ambipolar carbon nanotube transistors [15]. Anh et al. demonstrated that scanning photopresent microscopy was a valuable tool for imaging and optimizing electrical and optoelectronic properties of carbon nanotube based devices [15].
Dastagir et al. [16]	Dastagir et al. reported the unambiguous detection of a sequence of Hepatitis C Virus at concentrations down to the fractional pM range using Single Wall Carbon Nanotube Field Effect Transistor devices functionalized with Peptide Nucleic Acid [16].
Tarakanov and Kinaret [17]	Tarakanov and Kinaret investigated theoretically field effect transistors based on single-walled carbon nanotubes and explored two device geometries with suspended multiwalled carbon nanotubes functioning as gate electrodes [17]. Tarakanov and Kinaret suggested that the proposed designs can be used as nanoelectromechanical switches and as detectors of mechanical motion on the nanoscale [17].
Kauffman and Star [18]	Kauffman and Star demonstrated how the spectroscopic and electronic field-effect transistor measurements complimentary ways could be combined to further understand electronic modifications of the single-walled carbon nanotubes [18].
Fedorov et al. [19]	In this work, Fedorov et al. extract the quasi-metallic carbon nanotube chirality too as the characteristics of the Schottky barriers formed at the metal-nanotube contacts from the temperature-dependent magnetoconductance measurements [19].
Liang et al. [20]	In this work, field-effect transistors were fabricated using single-walled carbon nanotubes and double-walled carbon nanotubes, and their electrical transport properties were studied comparatively [20]. Liang et al. found that the semiconducting - metallic double-walled carbon nanotube devices thus bring a perfect system for studying the important intershell interaction [20].
Koswatta et al. [21]	Koswatte et al. performed extensive simulations for the band-to-band tunneling operation of carbon nanotube metal-oxide-semiconductor field-effect transistors using the nonequilibrium Green's function formalism for both ballistic and dissipative quantum transport [21]. Koswatta et al. obtained strong evidence that band-to-band tunneling in carbon nanotube metal-oxide-semiconductor field-effect transistors was dominated by optical phonon assisted inelastic transport, which can have important implications on the transistor characteristics [21]. Koswatta et al. mentioned that two-phonon scattering might also become important under large biasing conditions [21].

Table 2. (Continued)

Authors	Details
Wang et al. [22]	Wang et al. calculated the effect of single trapped charges in carbon nanotube field-effect transistors, using the nonequilibrium Greens function method in a tight-binding approximation. Wang et al. found that a single charge could shift and even rescale the entire transfer characteristic of the device [22].
Chen and Zhang [23]	Chen and Zhang studied carbon nanotube multi-channeled field-effect transistors. [23]. It was found that the transconductance of the multi-channel field-effect transistors has an almost linear dependency on the single-wall carbon nanotube channel number, which opens up a useful way to tune the transconductance of field-effect transistors by controlling the channel number [23].
Tersoff [24]	Tersoff revealed that nanoscale ballistic transistors gave voltage-dependent sensitivity to the intrinsic potential fluctuations from nearby charge traps [24].

B. Silicon Nanowire

Silicon nanowire is a commonly used nanomaterial. Because of its superior properties silicon carbide is one of the most useful materials for power electronics, hard- and biomaterials [25]. In the solid phase, the electronic and optical properties are controlled by the stacking of double layers of silicon and carbon atoms [25]. In thin films, a change in the stacking order often requires stress, which can be obtained naturally in systems with nanometre length scale [25]. Considering preparation of silicon nanowire, many vapor phase methods are developed for the preparation and modification of inorganic nanowires [26]. In 2005, Friedman et al presented the use of low-temperature processes to integrate high-characteristic multi-nanowire transistors into logical inverters and fast ring oscillators on glass substrates [27]. Besides potentially enabling powerful electronics to permeate all aspects of novel life, this advance could find applied usage in devices such as low-cost radio-frequency tags and completely integrated high-refresh-rate displays [27]. In addition, integrating nanophotonics with electronics could enhance and/or enable chances in areas ranging from communications and computing [28]. In 2006, Hayden et al reported avalanche multiplication of the photopresent in nanoscale p-n diodes consist ofing of crossed silicon-cadmium sulphide nanowires [28]. According to this work, crossed nanowire arrays revealed that nanowire avalanche photodiodes were reproducible and can be addressed independently without cross-talk and nanowire avalanche photodiodes and arrays could open new chances for ultradense integrated systems, sensing and imaging applied usages [28].

However, at present, there are some alternative computing methods that may bring commercial computing into the nano-realm without the use of transistors. The nanochips for nanocomputers are the best examples.

Table 3. Some interesting reports on carbon nanotubes for nanotransistors

Authors	Details
Elfstrom et al. [29]	The aim of this research was to analyze the size dependence of the sensitivity of such nanowires for biomolecule detection and for other sensor applied usages [29]. Elfstrom et al. noted that results from electrical characterization of the nanowires revealed a threshold voltage increasing with decreasing width [29].
Javey et al. [30]	Javey et al. report a general approach for three-dimensional (3D) multifunctional electronics based on the layer-by-layer assembly of nanowire building blocks [30]. Javey et al. said that the ability to assemble reproducibly sequential layers of distinct types of nanowire-based devices coupled with the breadth of nanowire building blocks should enable the assembly of increasing complex multilayer and multifunctional three-dimensional electronics in the future [30].
Winkelmann et al. [31]	Winkelmann et al. studied porphyrin derivative coated silicon nanowire field effect transistors (SiNW-FETs), which displayed a large, stable, and reproducible conductance increase upon illumination [31].
Huang et al. [32]	Huang et al. reported on nanowires for integrated multicolor nanophotonics [32]. In this work, nanoscale light-emitting diodes with colors spanning from the ultraviolet to near-infextremely rared region of the electromagnetic spectrum were treated using a solution-based approach in which emissive electron-doped semiconductor nanowires were integrated with nonemissive hole-doped silicon nanowires in crossed nanowire architecture [32]. Huang et al. said that this approach was generalized to hybrid devices consist ofing of nanowire emitters integrated on lithographically patterned planar silicon structures, which could bring a route for integrating photonic devices with conventional silicon microelectronics [32].
Weber et al. [33]	Weber et al. reported on silicon-nanowire transistors with intruded nickel-silicide contacts [33]. Weber et al. found that the gate length was shortened by the axial, self-aligned formation of nickel-silicide source and drain segments along the nanowire [33].
Goldberger et al. [34]	Goldberger et al. demonstrated the direct vertical integration of silicon nanowire arrays into surrounding gate field effect transistors without the need for postgrowth nanowire assembly processes [34]. They mentioned that the first-generation vertically integrated nanowire field effect transistors exhibited electronic properties that are comparable to other horizontal nanowire field effect transistors and might, with further optimization, compete with advanced solid-state nanoelectronic devices [34].
Patolsky et al. [35]	Patolsky et al. reported electrical properties of hybrid structures consisting of arrays of nanowire field-effect transistors integrated with the indivicouple axons and dendrites of live mammalian neurons, where each nanoscale junction could be used for spatially resolved, highly sensitive detection, stimulation, and/or inhibition of neuronal signal propagation [35].
Ahn et al. [36]	Ahn et al. reported optical scanning measurements on photopresent in indivicouple silicon nanowire field effect transistors [36]. Ahn et al. said that the magnitude and polarity of the photopresent varied depending on the gate bias, a property that could be explained using band flattening and a Schottky-barrier-type change [36].

NANOCOMPUTERS

As previously mentioned, the computer has become a new necessary scientific device that can help hasten the scientific process. Concept of downsizing has been applied in computer engineering for years. This can be seen in the continual miniaturization of the computer's most elementary component, the transistor. As transistors became smaller, more could be integrated into a single microchip, and so the computational power can be more than that of a previous version. However this miniaturization process is now reaching a limit, a quantum threshold below which transistors will stop their function. The well-known quantum phenomenon of electrons tunneling through the barriers between wires is the main obstacle [37 – 38]. In order to progress, an alternative to transistor technology was found, one whose components could retain function through quantum effects rather than in despite of them. Redesigning of the transistors into nanotransistors was performed and this could be the solution for producing nanocomputers in nanocomputing [39 – 43]. The first generation of nanocomputers, nanometre-scale classical computers, had components that behave according to quantum mechanics, but the algorithms that these computers ran would probably not involve quantum mechanics. After that the next generation that poses quantum mechanics-based algorithm was developed and it was called quantum computers.

Table 4. Some interesting reports on electron tunneling and nanocomputer

Authors	Details
Friedman and Averin [44]	Friedman and Averin reported on Aharonov-Casher-effect suppression of macroscopic tunneling of magnetic flux [44].
Fistul and Efetov [45]	Fistul and Efetov reported on electromagnetic-field-induced suppression of transport through n-p junctions in grapheme [45].
Ferguson et al. [46]	Ferguson et al. investigated the effect of Zeeman splitting on quasiparticle transport in normal-superconducting-normal (NSN) aluminum single-electron transistors [46]. In the above-gap transport, Ferguson et al. found that the interplay of Coulomb blockade and Zeeman splitting led to spin-dependence of the sequential tunneling [46].
Johannissen et al. [47]	Johannissen et al. revealed that, during proton tunneling in the oxidative deamination of tryptamine catalyzed by the enzyme aromatic amine dehydrogenase (AADH), a sub-picosecond promoting vibration was inherent to the iminoquinone intermediate [47].
Wensorra et al .[48]	Wensorra et al. reported on a quantum collimation effect based on surface depletion regions in AlAs/GaAs nanocolumns with an embedded resonant tunneling structure [48]. Wensorra et al. concluded that resonant tunneling in nanocolumns could be improved by quantum collimation [48].
Matsunaga and Sohlberg [49]	Matsunaga and Sohlberg presented a quantum mechanical tunneling model for electron transport through a molecular junction and addressed the model with the results of quantum chemical electronic structure calculations [49].

Quantum Computers

Quantum computer is the latest version of nanocomputer. Basically, a combination of laser light and phonons can induce non-trivial logic and can be used in quantum computation. This kind of quantum logic gates are presently being set by many experimental groups around the world. A major goal of producing high-precision control of many-body quantum systems is to realize their potential as quantum computers [50]. Basically, quantum computer requires the storage of quantum information in a set of two-level systems, which is called qubits, the processing of this information using quantum gates and a means of final output [51]. Assuming that a quantum computer could be constructed, it would in practice be required to function with noisy devices called "gates" [52]. These gates cause decoherence of the fragile quantum states that are central to the computer's operation [52]. Although quantum computation is probable if the noise impacting the quantum system satisfies certain conditions, existing methods for noise characterization are intractable for multibody systems [50]. However, only a few systems have been identified as potentially viable quantum computer models [51]. Examples include quantum optical systems (such as those utilizing trapped ions or neutral atoms, cavity quantum electrodynamics and nuclear magnetic resonance) and solid state systems (using nuclear spins, quantum dots and Josephson junctions) [51]. Some interesting reports on electron tunneling and quantum computer are presented in Table 5.

Table 5. Reports on electron tunneling and quantum computer

Authors	Details
Emerson et al. [50]	Emerson et al. introduced a way based on symmetrization that enables direct experimental measurement of some key properties of the decoherence impacting a quantum system [50]. Emerson et al. mentioned that the way was demonstrated for the optimization of control over nuclear spins in the solid state [50].
Cirac and Zoller [51]	Cirac and Zoller proposed a model for an ion trap quantum computer that combines scalability with the usefulness of quantum optical systems [51].
Knill [52]	Knill reported a simple architecture for fault-tolerant quantum computing, providing evidence that accurate quantum computing was probable for error probability per gate s as high as three per cent [52].
Kampermann and Veeman [53]	Characterization of quantum algorithms by quantum process tomography using quadrupolar spins in solid-state nuclear magnetic resonance was reported [53]. Kampermann and Veeman used the quadrupolar spin-32 sodium nuclei of a NaNO3 single crystal as a virtual two-qubit system [53].
Prevedel et al. [54]	Prevedel et al. realized a concatenated scheme of measurement and active feed-forward in a one-way quantum computing experiment and demonstrated that, for a perfect cluster state and no photon loss, their quantum computation scheme would operate with good fidelity and that their feed-forward components function with very high speed and low error for detected photons [54]. With this technology, Prevedel et al. proposed that the indivicouple computational step could be run in less than 150 nanoseconds using electro-optical modulators [54].

Table 5. (Continued)

Authors	Details
Viola et al. [55]	Experimental realization of noiseless subsystems for quantum information processing was performed [55]. Viola et al. demonstrated the protection of one bit of quantum information against all collective noise in three nuclear spins [55].
Khodjasteh and Lidar [56]	Khodjasteh and Lidar introduced a method of recursively concatenated dynamical decoupling pulses, designed to overcome both decoherence and operational errors [56]. Khodjasteh and Lidar derived a condition on the pulse noise level below which concatenation was guaranteed to reduce decoherence [56].
Kielpinski et al. [57]	Producing the array-based approach, Kielpinski et al. demonstrated how to get a massively parallel gate operation in a large-scale quantum computer, based on ways already demonstrated for manipulating small quantum registers [57]. Kielpinski et al. concluded that the use of decoherence-free subspaces significantly reduced decoherence during ion transport, and get rid ofs the requirement of clock synchronization between the interaction regions [57].
Hagan et al. [58]	Hagan et al. critically examined the decoherence mechanisms likely to dominate in a biological setting and found that topological quantum computation in microtubules might be error correcting, resistant to decoherence; and the decohering effect of radiative scatterers on microtubule quantum states was negligible [58].
Wu et al. [59]	Wu et al. revealed how to realize, by means of non-Abelian quantum holonomies, a set of universal quantum gates acting on decoherence-free subspaces and subsystems [59].
DiVincenzo et al. [60]	DiVincenzo et al. introduced an explicit scheme in which the Heisenberg interaction alone suffices to implement any quantum computer circuit [60]. DiVincenzo et al. said that this capability came at a price of a factor of three in additional qubits, and about a factor of ten in additional two-qubit operations [60].
Roos et al. [61]	Roos used a decoherence-free subspace with specifically designed entangled states to demonstrate precision spectroscopy of a pair of trapped calcium ions and obtained the electric quadrupole moment, which was of use for frequency standard applied usages [61]. Roos et al. found that entangled states were not only useful for enhancing the signal-to-noise ratio in frequency measurements [61].

Cognomics and Convergence [62 – 64]

At present, there are four main important branches of modern science, computer (science of the bit), biotechnology (science of the gene), nanotechnology (science of the atom) and networks (science of the neuron). Integration of seven technologies; diagnostic technology, tissue engineering, immunomodulation, genomics, therapeutic technology, medicine delivery technology and personal enhancement, can be seen. The bio-, nano-, info-, cogno-convergence will be the next step of science. The nanocomputer is an example of nano-cogno combination. Finally, in addition, to illuminate the societal implications of nano-bio-info-cogno technologies it is critical to place them within a broad historical context [62 – 64]. The plan for social context is also necessary [62 – 64]. Consistency by unity of theory and knowledge, theory connection hybrid of technologies and connection to society is the final destination.

Table 6. Convergence of four main branches of modern science

Combinations	Examples
Nano-bio	synthetic biology, RNA scaffolded nanostructures nanobiosensors, nanoparticles for sensing, delivery, nanobioarrays
Nano-info	single molecule transistors, nanowires, nanomemory, molecular dynamics, device optimization, ab initio calculations
Nano-cogno	force field optimization, artificial intelligence modeling of surface chemistry, nanoelectrodes nanocomputers
Bio-info	DNA nanotags, DNA computing, synthetic biology, structure prediction, pathway discovery, –omics analysis
Bio-cogno	automatic algorithm generation for bio-applied usages, bio-data mining, evolutionary computation, biorobots
Info-cogno	intelligent agents for network, agents for threat discovery, avatars and artificial intelligence, personal networking

REFERENCES

[1] Aeppli G, Pankhurst Q. Launch of the London Centre for Nanotechnology. *Nanomed.* 2006 Dec;1(4):393-8.

[2] Murphy D, Challacombe B, Nedas T, Elhage O, Althoefer K, Seneviratne L, Dasgupta P.Equipment and technology in robotics. *Arch. Esp. Urol.* 2007 May;60(4):349-55.

[3] Hamann DR, Muller DA. Absolute and approximate calculations of electron-energy-loss spectroscopy edge thresholds. *Phys. Rev. Lett.* 2002 Sep 16;89(12):126404.

[4] Appenzeller J, Radosavljevic M, Knoch J, Avouris P. Tunneling versus thermionic emission in one-dimensional semiconductors. *Phys. Rev. Lett.* 2004 Jan 30;92(4):048301.

[5] Jung SH, Jeong SH, Kim SU, Hwang SK, Lee PS, Lee KH, Ko JH, Bae E, Kang D, Park W, Oh H, Kim JJ, Kim H, Park CG. Vertically aligned carbon-nanotube arrays showing Schottky property at room temperature *Small.* 2005 May;1(5):553-9.

[6] Lu W, Lieber CM. Nanoelectronics from the bottom up. *Nat Mater.* 2007 Nov;6(11):841-50.

[7] Iijima S. Helical microtubules of graphitic carbon. S Iijima *Nature.* 1991; 354: 56.

[8] Praveena G, Kolandaivel P, Santhanamoorthi N, Renugopalakrishnan V, Ramakrishna S. Looking beyond carbon nanotubes: polypeptide nanotubes as alternatives? *J. Nanosci. Nanotechnol.* 2007 Jul;7(7):2253-9.

[9] Sinnott SB. Chemical functionalization of carbon nanotubes. *J. Nanosci. Nanotechnol.* 2002 Apr;2(2):113-23.

[10] Boczkowski J, Lanone S. Potential uses of carbon nanotubes in the medical field: how worried should patients be? *Nanomed.* 2007 Aug;2(4):407-10.

[11] Helland A, Wick P, Koehler A, Schmid K, Som C. Reviewing the environmental and human health knowledge base of carbon nanotubes. *Environ. Health Perspect.* 2007 Aug;115(8):1125-31.

[12] Koyama S, Haniu H, Osaka K, Koyama H, Kuroiwa N, Endo M, Kim YA, Hayashi T. Medical applied usage of carbon-nanotube-filled nanocomposites: the microcatheter. *Small.* 2006 Dec;2(12):1406-11.

[13] Javey A, Guo J, Wang Q, Lundstrom M, Dai H. Ballistic carbon nanotube field-effect transistors. *Nature.* 2003 Aug 7;424(6949):654-7.

[14] Gui EL, Li LJ, Zhang K, Xu Y, Dong X, Ho X, Lee PS, Kasim J, Shen ZX, Rogers JA, Mhaisalkar SG. DNA Sensing by Field-Effect Transistors Based on Networks of Carbon Nanotubes. *J. Am. Chem. Soc.* 2007 Oct 31;

[15] Ahn YH, Tsen AW, Kim B, Park YW, Park J. Photopresent Imaging of p-n Junctions in Ambipolar Carbon Nanotube Transistors. *Nano. Lett.* 2007 Oct 16;

[16] Dastagir T, Forzani ES, Zhang R, Amlani I, Nagahara LA, Tsui R, Tao N. Electrical detection of hepatitis C virus RNA on single wall carbon nanotube-field effect transistors. *Analyst.* 2007 Aug;132(8):738-40.

[17] Tarakanov YA, Kinaret JM. A carbon nanotube field effect transistor with a suspended nanotube gate. *Nano Lett.* 2007 Aug;7(8):2291-4.

[18] Kauffman DR, Star A. Single-walled carbon-nanotube spectroscopic and electronic field-effect transistor measurements: a combined approach. *Small.* 2007 Aug;3(8):1324-9.

[19] Fedorov G, Tselev A, Jimenez D, Latil S, Kalugin NG, Barbara P, Smirnov D, Roche S. Magnetically induced field effect in carbon nanotube devices. *Nano Lett.* 2007 Apr;7(4):960-4.

[20] Fedorov G, Tselev A, Jimenez D, Latil S, Kalugin NG, Barbara P, Smirnov D, Roche S. Magnetically induced field effect in carbon nanotube devices. *Nano Lett.* 2007 Apr;7(4):960-4.

[21] Koswatta SO, Lundstrom MS, Nikonov DE. Band-to-band tunneling in a carbon nanotube metal-oxide-semiconductor field-effect transistor is dominated by phonon-assisted tunneling. *Nano Lett.* 2007 May;7(5):1160-4.

[22] Wang NP, Heinze S, Tersoff J. Random-telegraph-signal noise and device variability in ballistic nanotube transistors. *Nano Lett.* 2007 Apr;7(4):910-3.

[23] Chen C, Zhang Y. Carbon nanotube multi-channeled field-effect transistors. *J. Nanosci. Nanotechnol.* 2006 Dec;6(12):3789-93.

[24] Tersoff J. Low-frequency noise in nanoscale ballistic transistors. *Nano Lett.* 2007 Jan;7(1):194-8.

[25] Melinon P, Masenelli B, Tournus F, Perez A. Playing with carbon and silicon at the nanoscale. *Nat Mater.* 2007 Jul;6(7):479-90.

[26] Govindaraj A, Vivekchand SR, Rao CN. New vapor phase reactions for the synthesis and modification of carbon nanotubes and inorganic nanowires. *J. Nanosci. Nanotechnol.* 2007 Jun;7(6):1695-702.

[27] Friedman RS, McAlpine MC, Ricketts DS, Ham D, Lieber CM. Nanotechnology: high-speed integrated nanowire circuits. *Nature.* 2005 Apr 28;434(7037):1085.

[28] Hayden O, Agarwal R, Lieber CM. Nanoscale avalanche photodiodes for highly sensitive and spatially resolved photon detection. *Nat. Mater.* 2006 May;5(5):352-6.

[29] Elfstrom N, Juhasz R, Sychugov I, Engfeldt T, Karlstrom AE, Linnros J. Surface charge sensitivity of silicon nanowires: size dependence. *Nano Lett.* 2007 Sep;7(9):2608-12

[30] Javey A, Nam S, Friedman RS, Yan H, Lieber CM. Layer-by-layer assembly of nanowires for three-dimensional, multifunctional electronics. *Nano Lett.* 2007 Mar;7(3):773-7.

[31] Winkelmann CB, Ionica I, Chevalier X, Royal G, Bucher C, Bouchiat V. Optical switching of porphyrin-coated silicon nanowire field effect transistors. *Nano Lett.* 2007 Jun;7(6):1454-8.

[32] Huang Y, Duan X, Lieber CM. Nanowires for integrated multicolor nanophotonics. *Small.* 2005 Jan;1(1):142-7.

[33] Weber WM, Geelhaar L, Graham AP, Unger E, Duesberg GS, Liebau M, Pamler W, Cheze C, Riechert H, Lugli P, Kreupl F. Silicon-nanowire transistors with intruded nickel-silicide contacts. *Nano Lett.* 2006 Dec;6(12):2660-6.

[34] Goldberger J, Hochbaum AI, Fan R, Yang P. Silicon vertically integrated nanowire field effect transistors. *Nano Lett.* 2006 May;6(5):973-7.

[35] Patolsky F, Timko BP, Yu G, Fang Y, Greytak AB, Zheng G, Lieber CM. Detection, stimulation, and inhibition of neuronal signals with high-density nanowire transistor arrays. *Science.* 2006 Aug 25;313(5790):1100-4.

[36] Ahn Y, Dunning J, Park J. Scanning photopresent imaging and electronic band studies in silicon nanowire field effect transistors. *Nano Lett.* 2005 Jul;5(7):1367-70.

[37] Wang KL. Issues of nanoelectronics: a probable roadmap. *J. Nanosci. Nanotechnol.* 2002 Jun-Aug;2(3-4):235-66.

[38] Selzer Y, Allara DL. Single-molecule electrical junctions. *Annu. Rev. Phys. Chem.* 2006;57:593-623.

[39] Normile D. Nanocomputing. The end--not here yet, but coming soon. *Science.* 2001 Aug 3;293(5531):787.

[40] Service RF. Nanocomputing. World's smallest transistor. *Science.* 2001 Aug 3;293(5531):786

[41] Service RF. Nanocomputing. Optical lithography goes to extremes--and beyond. *Science.* 2001 Aug 3;293(5531):785-6.

[42] Normile D, Service RF. Nanocomputing. Yet another role for DNA? *Science.* 2001 Aug 3;293(5531):783.

[43] Service RF. Nanocomputing. Assembling nanocircuits from the bottom up. *Science.* 2001 Aug 3;293(5531):782-5.

[44] Friedman JR, Averin DV. Aharonov-Casher-effect suppression of macroscopic tunneling of magnetic flux. *Phys. Rev. Lett.* 2002 Feb 4;88(5):050403.

[45] Fistul MV, Efetov KB. Electromagnetic-field-induced suppression of transport through n-p junctions in graphene. *Phys. Rev. Lett.* 2007 Jun 22;98(25):256803.

[46] Ferguson AJ, Andresen SE, Brenner R, Clark RG. Spin-dependent quasiparticle transport in aluminum single-electron transistors. *Phys. Rev. Lett.* 2006 Aug 25;97(8):086602.

[47] Johannissen LO, Hay S, Scrutton NS, Sutcliffe MJ. Proton tunneling in aromatic amine dehydrogenase is driven by a short-range sub-picosecond promoting vibration: consist ofency of simulation and theory with experiment. *J. Phys. Chem. B.* 2007 Mar 15;111(10):2631-8.

[48] Wensorra J, Indlekofer KM, Lepsa MI, Forster A, Luth H.Resonant tunneling in nanocolumns improved by quantum collimation. *Nano Lett.* 2005 Dec;5(12):2470-5.
[49] Wang KL. The substituent effect on nonlinear present voltage characteristics in a two-terminal molecular electronic device. *J. Nanosci. Nanotechnol.* 2001 Sep;1(3):275-9.
[50] Emerson J, Silva M, Moussa O, Ryan C, Laforest M, Baugh J, Cory DG, Laflamme R. Symmetrized characterization of noisy quantum processes. *Science.* 2007 Sep 28;317(5846):1893-6.
[51] Cirac JI, Zoller P. A scalable quantum computer with ions in an array of microtraps *Nature.* 2000 Apr 6;404(6778):579-81.
[52] Knill E. Quantum computing with realistically noisy devices. *Nature.* 2005 Mar 3;434(7029):39-44.
[53] Kampermann H, Veeman WS. Characterization of quantum algorithms by quantum process tomography using quadrupolar spins in solid-state nuclear magnetic resonance. *J. Chem. Phys.* 2005 Jun 1;122(21):214108.
[54] Prevedel R, Walther P, Tiefenbacher F, Bohi P, Kaltenbaek R, Jennewein T, Zeilinger A. high-speed linear optics quantum computing using active feed-forward. *Nature.* 2007 Jan 4;445(7123):65-9.
[55] Viola L, Fortunato EM, Pravia MA, Knill E, Laflamme R, Cory DG. Experimental realization of noiseless subsystems for quantum information processing. *Science.* 2001 Sep 14;293(5537):2059-63.
[56] Khodjasteh K, Lidar DA. Fault-tolerant quantum dynamical decoupling. *Phys Rev. Lett.* 2005 Oct 28;95(18):180501.
[57] Kielpinski D, Monroe C, Wineland DJ. Architecture for a large-scale ion-trap quantum computer. *Nature.* 2002 Jun 13;417(6890):709-11.
[58] Hagan S, Hameroff SR, Tuszynski JA. Quantum computation in brain microtubules: decoherence and biological feasibility. *Phys. Rev. E Stat. Nonlin. Soft Matter. Phys.* 2002 Jun;65(6 Pt 1):061901.
[59] Wu LA, Zanardi P, Lidar DA. Holonomic quantum computation in decoherence-free subspaces. *Phys. Rev. Lett.* 2005 Sep 23;95(13):130501.
[60] DiVincenzo DP, Bacon D, Kempe J, Burkard G, Whaley KB. Universal quantum computation with the exchange interaction. *Nature.* 2000 Nov 16;408(6810):339-42.
[61] Roos CF, Chwalla M, Kim K, Riebe M, Blatt R. 'Designer atoms' for quantum metrology. *Nature.* 2006 Sep 21;443(7109):316-9.
[62] Lynch Z. Neurotechnology and society (2010-2060). *Ann. N. Y. Acad. Sci.* 2004 May;1013:229-33.
[63] Khushf G. Systems theory and the ethics of human enhancement: a framework for NBIC convergence. *Ann. N .Y. Acad. Sci.* 2004 May;1013:124-49.
[64] Hochella MF Jr. The case for nanogeoscience. *Ann. N. Y. Acad. Sci.* 2006 Dec;1093:108-22.

Chapter 12

NANOQUANTITY MEASUREMENT IN MEDICINE

INTRODUCTION TO NANOQUANTITY IN MEDICINE

Often, the measurement of quantity in medicine results in a nanolevel summation. Although those substances are very few, they still exist. Many nanoquantity substances in human beings have a significant biological effect. A good example of nanoquantity substances in the human body are vitamins, cardiac markers, tumor markers, and hormones. The measurement of these nanoquantity substances is an important concern in laboratory medicine. Because of the nanoquantity, the precision and accuracy of the analytical system is very important.

Table 1. Concerned parameters for nanoquantity measurement in medicine

Parameter	Details
1. Analytical characteristic	
1.1 Analytical sensitivity	The lower detection limit of the test
1.2 Linearity	The ability of a test to obtain results that vary in a manner directly proportional to changes in the summation of the analyte in the sample
1.3 Precision	The ability of a measurement to be consist ofently reproduced.
1.4 Accuracy	The percentage of true correct results
2. Clinical characteristic	
2.1 Sensitivity	The proportion of true positives it detects of all the positives
2.2 Specificity	The proportion of true negatives it detects of all the negatives
2.3 Positive predictive value	Percent of patients with positive test having disease
2.4 Negative predictive value	Percent of patients with negative test not having disease

In this chapter, important nanoquantity measurement methods in medicine will be further described and discussed.

ELECTROCHEMILUMINESCENCE

1. Principle of Electrochemiluminescence

Electrochemiluminescence or electrogenerated chemiluminescence is a form of chemiluminescence in which the light-emitting chemiluminescent reaction is preceded by an electrochemical reaction [10 -11]. The usefulness of chemiluminescence are retained, but the electrochemical reaction allows the time and position of the light emitting reaction to be controlled. By controlling the time light emission can be delayed until events such as immune or enzyme catalyzed reactions have taken place [10 -11]. For a mechanism, the combination of electrochemiluminescence with universally coated micro particle technologies results in fast assays with very high sensitivity and a broad measuring range [10 -11]. This is obtained by optimised selection of the electrochemically active "substrates" (Ruthenium chelates and tris-propylamine, TPA) [12]. First, the substrate will reach the cathode to obtain in the excessive electron and becomes negatively charged. Second, the analyte will reach the anode to give out an electron and becomes positively charged. Third, substrate and analyte meet each other. The transmission of the excessive electron from substrate to the electron-depleted analyte occurs. The electron transmission is based on the principle of electron valency change. Finally, the transmission of an electron causes the transmission of a photon that can be recorded and interpreted as quantity of analyte. The use of a confocal microspectrometer allows in situ photon detection of the eleectrochemiluminescence reaction with micrometric spatial resolution and observes the concentration distribution of the photon source in the vicinity of the interelectrode gap [13]. Electrochemiluminescence is presently used as an advanced way for new clinical chemistry analyzer for microanalyte including cardiac markers, tumor markers and hormones.

2. Determination of Hormone by Electrochemiluminescence

2.1. Determination of Sex Hormone by Electrochemiluminescence

Determination of sex hormone can be performed by electrochemiluminescence. There are some interesting recent reports on the determination of sex hormone by electrochemiluminescence.

A. Study on Analytical Characteristic

Yang et al. performed a comparative study of estradiol analysis by many analyzers and they found that estradiol assay by electrochemiluminescence method was acceptable and accurate [14]. Gassler et al. established reference ranges for lutropin, follitropin, estradiol, testosterone, and prolactin by electrochemiluminescence immunoassays (Elecsys 2010, Roche Diagnostics) in 299 children of eight different age groups [15]. They found that the estradiol concentrations did not differ between the genders but dropped significantly in the first few months of life and the testosterone concentrations were higher in male than in female children, whereas the concentrations of follitropin were higher in female than in male children [15].

B.Study on Clinical Applied Usage

Kopp et al studied a probable interrelationship between androgens, insulin resistance and inflammation in a prospective study with 43 morbidly obese female patients undergoing bariatric surgery and found that massive weight loss induced by bariatric surgery was associated with a significant reduction of androgen concentrations [16]. Erbagci et al recently studied the menstrual cycle dependent variability for serum tumor markers CEA, AFP, CA 19-9, CA 125 and CA 15-3 in healthy women [17]. In this study midfollicular and midluteal concentrations of serum CEA, AFP, CA 19-9, CA 125, CA 15-3 and their relations with LH, FSH, prolactin, estradiol and progesterone were assessd during ovulatory cycles in a group of 23 healthy female indivicouples [17]. They suggested that in healthy women serum CA 15-3 levels are significantly elevated in the midluteal phase of the menstrual cycle compared to midfollicular phase, therefore, consideration of menstrual cycle dependent variability for CA 15-3 appears indicated in interpretation of indivicouple results [17]. Wiwanitkit also recently concluded that electrochemiluminescence is the useful diagnostic tool in sexual medicine [18].

2.2. Determination of Thyroid Hormone by Electrochemiluminescence

Determination of thyroid hormone can be performed by electrochemiluminescence. There are some interesting recent reports on the determination of thyroid hormone by electrochemiluminescence.

A. Study on Analytical Characteristic

In 1998, Forest et al assessd the analytical characteristic of the Elecsys 2010 immunoanalyzer, which was based on a new detection technology, electrochemical luminescence [19]. According to this work, the Elecsys 2010 system was shown to have an acceptable analytical characteristic for the rapid analysis of a wide multiple appearance of analytes [19]. In addition, the hook effect observed with hCG assay would imply that the laboratory informs the clinicians of the possibility of falsely low values in trophoblastic diseases or that all specimens with values greater than 3000 U/L were reassayed after dilution [19]. Sanchez-Carbayo et al performed another analytical evaluation of the new electrochemiluminescent immunoassays for TSH, FT4, and T3 in the Elecsys 2010 immunoassay system [20]. Sanchez-Carbayo et al concluded that the Elecsys 2010 was especially attractive as a routine assay because it was completely automated, obtaining results in only 18 minutes [20]. They reported that the analytical assay characteristic for TSH, FT4 and T3 was shown to be acceptable [20]. In addition, they noted that using two different sets of diagnostic tests, minimal discrepancies were found in the laboratory assessment for the classification of patients with clinical suspicion of thyroid disease [20]. In 1998, Luppa et al presented the results of a pre-evaluation of the thyroid function test FT4, FT3 and third generation TSH using the Elecsys electrochemiluminescence immunoassay system [21]. The aim of this field study was to support the optimization of the technology used for Elecsys in an early stage of development and thereby prepare the ground for the adaptation of the immunoassays to the final Elecsys 2010 random access analyzer [21]. Luppa et al demonstrated that the requirements of routine thyroid testing in terms of reliability were fulfilled by the system [21]. Characteristic of the electrochemiluminescent immunoassay system of Elecsys 2010 was also assessd by quantitative assay by Qin [22]. According to this work, the functional sensitivity of the TSH assay was 0.004 microU/ml [22]. Qin concluded that the Elecsys 2010 was provably satisfactory in sensitivity, precision, accuracy, reliability, practicability, and rapidity [22]. Similar results were also reported in the study of Sapin et al [23].

B.Study on Clinical Applied Usage

Semlitsch et al examined thyroid function in six patients with known T3-binding autoantibodies using a radioimmunoassay and an electrochemiluminescence immunoassay [24]. They concluded hat in patients with triiodothyronine-binding autoantibodies FT3 measurement by electrochemiluminescence way was more useful than measurement by radioimmunoassay [24].

3. Determination of Tumor Marker by Electrochemiluminescence

Tumor marker is a group of laboratory investigations. Molecules occurring in the human body associated with cancer are tumor markers. Tumor marker can be used for diagnosis or management of disease. Ideal marker is a "blood test" that is positive only when patient has the disease. In clinical pathology, tumor marker poses a very long history. The first tumor marker is Bence Jones protein, a urine protein that was firstly described by Bence Jones in 1847 [25]. This protein is the tumor marker for multiple myeloma, cancer of plasma cell. The other classical tumors include acid phosphatase in advanced prostate cancer which was discovered by Gutman and Gutman in 1938 [26]. While oncofetal antigens and prostate specific antigen are the examples of new tumor markers discovered in the late twentieth century. Five groups of serum tumor markers are available at present [27]. The enzyme and hormone are not specific since these two groups can be secreted from the normal cells. Of five groups, tumor antigens are most specific. Two subgroups of tumor antigens, oncofetal and non-oncofetal proteins are mentioned [27]. Basically, the protein synthesis starts within the nucleus. DNA translates via mRNA into protein in cytosol and secrete out to the membrane and transport through cell membranes into the external environment. These proteins can be then detected in serum. Any abnormal proteins due to the carcinogenesis process can be the candidate for a tumor marker test. Monoclonal antibodies can be developed and used to detect these proteins on the cancer cell surface or in the serum. The five main uses for tumor markers are screening, diagnosis, monitoring therapy, monitoring remission, and following-up.

3.1. Some Problems with the Present Tumor Markers Investigation in Clinical Pathology

3.1.1. Lack of Good Test

Screening and diagnosis are both focused properties of good tumor markers. For screening, high sensitivity is needed. While for diagnosis, high specificity is needed. Any test with these two combined properties are difficult to find. Of many tumor markers, PSA is the best marker in clinical pathology [28].

Most of the present tumor markers lack specificity due to the high false negatives. Benign diseases positive CA 125 [29] or CEA [30] are common. Some physiological underlying can impact the analytical result such as smokers have a increased CEA [31]. Also, cancer heterogeneity usually exists for many organs [32]. In addition, many cases die with, not from, cancers.

3.1.2. How Can We Manage the Tumor Marker Test?

Due to the preventive medicine principle, early diagnosis is good preventive action. Screening is a generalized used benefit of tumor markers. However, there are many problems such as "Is a negative reassuring?" and "What does a positive indicate?". Management of the anxiety due to the test result is a necessary point [33 - 34]. Sometimes, the increased level of tumor markers can be seen in benign diseases. On the other hand, some increase can be detected in the late stage, unresectable mass. These facts are the flaws of the tumor markers screening.

For the follow up, a fall after therapy usually means a response, however, "normal" does not mean remission [32]. For the query: "Is palliative therapy is treating the symptom or the tumor marker level," us of the laboratory results accompanied with the signs and symptoms is suggested. However, it should be noted that not all symptoms are cancer progression. At the end of therapy, the physician should beware of misleading. Normal level does not mean cure, does not mean complete remission, may not even mean improvement. This can create anxiety, false reassurance and false hope [32].

3.1.3. Communication Gaps between the Patients and Physician

"Does the patient understand?" is the main question when a physician in charge orders the tumor marker testing. Explanation is necessary. Creating of anxiety has to be managed. For the oncologist, interpretation of the test result should be carecompletely performed. Sometimes, with multiple markers elevated and another marker falling, an increasing mass can be seen. In addition, the oncologist should be accessible to the family. Physicians who know the patient best, must answer the questions "why?" or "what next"? from the family.

3.2. Study on Analytica land Clinical Characteristic

A. Study on Analytical Characteristic

Stockmann et al. performed an evaluation on the Elecsys 2010 system for determination of PSA [35]. According to this study, the Elecsys 2010 system offered a high analytical sensitivity in combination with relatively low sample volumes and short measuring times; a high dynamic measuring range reduces the number of reruns and makes the work-flow more convenient [35]. Stockmann et al. reported that the electrochemiluminescence technology of Elecsys 2010 revealed usefulness in system characteristic without having any drawbacks compared to existing technologies [35].The CA 125 II assay on the Elecsys(R) 2010 analyzer was also assessd in an international multicenter experiment by van Ingel et al [36]. van Ingen said that the Elecsys CA 125 II assay was linear over a broad range, yielded precise and accurate results, was free from interferences, and compares well with other assays [36]. A similar study was also performed by Hubl et al. [37]. According to this work, the maximum of diagnostic efficiency of ovarian carcinoma patients compared with patients of benign gynecological diseases was reached at 150 U/ml with a specificity of 93% and a sensitivity of 69% [37]. Hubl et al. noted that the technical and clinical evaluation of the Elecsys CA 125 II assay revealed a superior analytical characteristic with a broad measuring range up to 5000 U/ml and a short measuring time of 18 minutes [37]. Stieber et al. performed an evaluation of the analytical and clinical characteristic of the Elecsys CA 15-3 immunoassay [38]. Stern et al. compared eight routinely employed immunoassay sets for the determination of MUC1 mucins in the serum: six closed automated systems (AxSYM, Centaur, ECi Vitros, Elecsys 2010, Immulite 2000 and Kryptor), and two IRMA kits (ELSA CIS and IRMA-mat Byk-Sangtec) [39]. In this work, the sensitivity was highest for the Elecsys 2010 [39]. Stern et al. also compared six routinely employed immunoassay kits: Architect i2000 and AxSYM, Abbott Laboratories; Elecsys 2010, Roche Diagnostics; ELSA, CIS-BioInternational; Immulite 1, Diagnostic Products Corporation; and IRMA-mat, Byk-Sangtec Diagnostica for determination of CA19-9 [40]. Stern et al. stated that the results of determining CA 19-9 could not be extrapolated from one analytical way to another, even in cases where the same monoclonal antibody was used [40]. Stern et al. noted that standardization of CA 19-9 measurement systems was necessary to allow use of the results for the purposes of evidence-based medicine [40]. According to this fact, Passerini et al. indicated that during tumor follow-up, the use of the same system was fit to avoid the risk of a variation due to the method rather than the disease [41]. Moreover, Passerini et al. also noted that whenever a change in analytical equipment was required, careful analysis of CA 19-9 results had to be undertaken [41]. Sanchez-Carbayo et al. performed an analytical evaluation of the CYFRA 21-1 electrochemiluminescence assay for serum and urine samples [42].

Sanchez-Carbayo et al. revealed reliable analytical characteristic for urinary CYFRA 21-1, which might assist urologists in the detection of bladder cancer as a noninvasive addition to cystoscopy [42].

B. Study on Clinical Characteristic

Filella et al. determined the serum levels of PSA, fPSA and proPSA using automated electrochemiluminescent immunoassays (Elecsys 2010, Roche Diagnostics) in the patients with prostate cancer and the patients with benign prostate hyperplasia [43]. According to this work, it was detected that the combination of the proPSA concentration and the proPSA/total PSA ratio improved the area under the curve obtained for indivicouple tests [43]. Fillela et al. said that ProPSA might be useful in the diagnosis of prostate cancer [43]. However, Lein et al. found that no improvement in diagnostic accuracy was shown between cPSA with tPSA or the ratio of f/tPSA in the tPSA range of less than 4 ng/ml [44]. Gray et al. studied comparability of serum PSA measurement between the Roche Diagnostics Elecsys 2010 and the Abbott Architect i2000 [45]. Gray et al. noted that because the results from the two assays were different, reference intervals appropriate to the method of PSA measurement should be used [45]. Vollmer et al. studied serum CYFRA 21-1 in advanced stage non-small cell lung cancer (NSCLC) [46]. Vollmer et al. reported that the initial level of serum CYFRA appeared to bring more prognostic information than clinical stage in advanced stage NSCLC [46]. They also reported that a drop of more than 27% in CYFRA after one cycle of therapy added prognostic information, so that this threshold appeared to be an early measure of response to chemotherapy [46]. Mulley et al. performed another study to assess the technical characteristic of the new double-monoclonal, single-step Elecsys neuron-specific enolase (NSE) enzyme immunoassay (EIA) [47]. Muley et al. noted that Elecsys NSE EIA was a reliable and accurate diagnostic procedure for the measurement of NSE in serum samples. Muley et al. also noted that the special merits of this assay were the wide measuring range and a short incubation time of 18 minutes [47].

4. Determination of Vitamin by Electrochemiluminescence

Owen and Roberts performed a comparison of different automated serum and whole blood folate assays (Advia Centaur, ARCHITECT i2000, Elecsys 2010, and IMMULITE 2000 methods) [48]. According to this study, the ARCHITECT i2000 folate assay demonstrated the best analytic characteristic and the poor agreement seen with whole blood hemolysates likely was due to calibration differences and differences in hemolysate preparation conditions [48]. Difference in results from a different analyzer was also quoted. However, this result is not concordant with that of Vogeser and Lorenzl which indicated the comparability of vitamin B12 level results between different analyzers [49]. Nevertheless, Snyder et al. also reported a similar interference in their study [50]. Snyder et al. recommended that laboratories conducted hemolysis interference studies for all laboratory test protocols [50].

5. Determination of Cardiac Marker by Electrochemiluminescence

A. Study on Analytical Characteristic

The analytical property of cardiac marker measurement by electrochemiluminescence analyzer was warranted by Forest et al [51]. Method comparison of cardiac marker assays on PATHFAST, StratusCS, AxSYM, Immulite 2000, triage, elecsys and cardiac reader was performed by Peetz et al. [52]. Peetz et al. reported that this evaluation study warranted the total good correlation of the results obtained with assays for cardiac markers developed on the PATHFAST analyzer with those on other immunoassay platforms and thus the analytical reliability of the developed methods [52]. Hetland and Dickstein also compared the cardiac troponin T by Elecsys system and a rapid ELISA [53]. Hetland and Dickstein found good agreement of the analytical results [53].

B. Study on Clinical Characteristic

Bionda et al. performed an interesting study to assess the clinical characteristic of 2 troponins assays: Troponin Ic 2(nd) generation (AccuTnI) on Access 2 of Beckman Coulter and Troponin Tc 3(rd) generation (Troponin T STAT) on Elecsys 2010 of Roche Diagnostics [54]. The results indicated that Beckman Coulter for cardiac troponin T was superior to the others [54]. According to this work, a few false positive and negative cardiac troponin T values had been observed in patients with complex pathologies [54]. Bionda et al. noted that this eventuality had to be taken into consideration if clinical findings were not in good accordance with laboratory results [54]. However, Heeschen et al. said that despite the lower cardiac specificity for cardiac troponin T, it appeared to have a stronger association with the patients' outcome [55].

Radioimmunoassay

Radioimmunoassay is another technique in medicine that can be used for determination of nanoquantity substances in the human body. The development of radioimmunoassay by

Berson and Yalow during the late 1950s represents a milestone in the history of the applied usage of radionuclide methodology to biology and to medical investigation and practice [56 - 57]. Radioimmunoassay is based upon the competition between labeled and unlabeled antigen for specific antibody sites, forming antigen-antibody complexes [56 - 62]. At equilibirum, the radioactive complex (bound) is separated from the radioactive antigen (free) [56 - 62]. The bound/free ratio is dependent upon the summation of nonradioactive antigen and antigen concentration in unknown samples is detected by comparing the bound/free ratio to the bound/free ratios obtained by incubating varying summations of known nonradioactive antigen with the same summation of antibody as in the unknown sample under similar assay conditions [56 - 62]. Similar to electrochemiluminescence way, radioimmunoassay can be used for determination of vitamins and hormones [63 – 69]. However, radioimmunoassay is based on the usage of a radioactive substance, therefore, its use is limited to avoid the contact of the radioactive substance by the medical technologist. The electrochemiluminescence which is safer has become commonly used instead of the old radioimmunoassay way.

REFERENCES

[1] Tanaka K, Toriumi M, Shinohara Y, Okazaki R. Disorders of gonadal function in hypothalamic-pituitary diseases. *Nippon Rinsho.* 1997;55:2969-73.

[2] Marshall JC. Clinics in endocrinology and metabolism. Investigative procedures. *Clin. Endocrinol. Metab.* 1975;4:545-67.

[3] Cortes-Gallegos V. Clinical and functional correlation of plasma steroids--diagnostic uses. *J. Steroid. Biochem.* 1975;6:689-94.

[4] Cortes-Gallegos V. A simple procedure for the combined determination of plasma estrogen and androgen concentrations by competitive protein binding analysis. *J. Steroid. Biochem.* 1972;3:725-33.

[5] Horgan ED, Riley WJ. Protein-binding assay for plasma testosterone, after purification by column partition chromatography. *Clin. Chem.* 1974;20:430-5.

[6] Tulchinsky D, Abraham GE. Radioimmunoassay of plasma estriol. *J. Clin. Endocrinol. Metab.* 1971;33:775-82.

[7] Elder PA, Lewis JG. An enzyme-linked immunosorbent assay (ELISA) for plasma testosterone. *J. Steroid. Biochem.* 1985;22:635-8.

[8] Elder PA, Yeo KH, Lewis JG, Clifford JK. An enzyme-linked immunosorbent assay (ELISA) for plasma progesterone: immobilised antigen approach. *Clin. Chim. Acta.* 1987;162:199-206.

[9] Preti MS, Lodi S, Busacchi P, Filicori M, Flamigni C. ELISA for salivary and plasma estriol in pregnancy. *Steroids.* 1984;43:469-79.

[10] Kissinger PT. Electrochemical detection in bioanalysis. *J. Pharm. Biomed. Anal.* 1996;14:871-80.

[11] Richter MM. Electrochemiluminescence (ECL). *Chem. Rev.* 2004;104:3003-36.

[12] Bruce D, McCall J, Richter MM. Effects of electron withdrawing and donating groups on the efficiency of tris(2,2'-bipyridyl)ruthenium(II)/tri-n-propylamine electrochemi-luminescence. *Analyst.* 2002;127:125-8.

[13] Amatore C, Pebay C, Servant L, Sojic N, Szunerits S, Thouin L. Mapping electrochemiluminescence as generated at double-band microelectrodes by confocal microscopy under steady state. *Chemphyschem.* 2006;7:1322-7.

[14] Yang DT, Owen WE, Ramsay CS, Xie H, Roberts WL. Characteristic characteristics of eight estradiol immunoassays. *Am. J. Clin. Pathol.* 2004;122:332-7.

[15] Gassler N, Peuschel T, Pankau R. Pediatric reference values of estradiol, testosterone, lutropin, follitropin and prolactin. *Clin. Lab.* 2000;46:553-60.

[16] Kopp HP, Krzyzanowska K, Schernthaner GH, Kriwanek S, Schernthaner G. Relationship of Androgens to Insulin Resistance and Chronic Inflammation in Morbidly Obese Premenopausal Women: Studies before and after Vertical Banded Gastroplasty. *Obes. Surg.* 2006;16:1214-20.

[17] Wiwanitkit V. Advanced way for sex hormone determination by electrochemiluminescence and applied usage to disability research. *Sex Disabil.* 2007 Sep; 25(3): 141 -6.

[18] Erbagci AB, Yilmaz N, Kutlar I. Menstrual cycle dependent variability for serum tumor markers CEA, AFP, CA 19-9, CA 125 and CA 15-3 in healthy women. *Dis. Markers.* 1999;15:259-67.

[19] Forest JC, Masse J, Lane A. Evaluation of the analytical characteristic of the Boehringer Mannheim Elecsys 2010 immunoanalyzer. *Clin. Biochem.* 1998 Mar;31(2):81-8.

[20] Sanchez-Carbayo M, Mauri M, Alfayate R, Miralles C, Soria F. Analytical and clinical evaluation of TSH and thyroid hormones by electrochemiluminescent immunoassays. *Clin. Biochem.* 1999 Aug;32(6):395-403.

[21] Luppa PB, Reutemann S, Huber U, Hoermann R, Poertl S, Kraiss S, von Bulow S, Neumeier D. Pre-evaluation and system optimization of the Elecsys thyroid electrochemiluminescence immunoassays. *Clin. Chem. Lab. Med.* 1998 Oct;36(10):789-96.

[22] Qin W. Evaluation of analytical methods and workflow characteristic of the Architect ci8200 integrated serum/plasma analyzer system. *Clin. Chim. Acta.* 2005 Jul 1;357(1):43-54.

[23] Sapin R, Gasser F, d'Herbomez M, Wemeau JL, Ebert C, Schlienger JL. Elecsys thyrotropin (TSH) assay assessd. *Clin. Chem.* 1997 Mar;43(3):545-7.

[24] Semlitsch G, Buchinger W, Reiterer E, Binter G, Rainer F. Measurement of free triiodothyronine (FT3) using an electrochemiluminescence immunoassay in patients with autoantibodies to triiodothyronine. *Acta Med. Austriaca.* 2000;27(2):54-5.

[25] Stone MJ. Henry Bence Jones and his protein. J Med Biogr. 1998 Feb;6(1):53-7.

[26] Sproul EE. Acid phosphatase and prostate cancer: historical overview. *Prostate.* 1980;1(4):411-3.

[27] Ochi Y, Okabe H, Inui T, Yamashiro K. Tumor marker--present and future. *Rinsho Byori.* 1997 Sep;45(9):875-83.

[28] Frydenberg M, Wijesinha S. Diagnosing prostate cancer - what GPs need to know. *Aust Fam Physician.* 2007 May;36(5):345-7.

[29] Devine PL, McGuckin MA, Ward BG. Circulating mucins as tumor markers in ovarian cancer (review). *Anticancer Res.* 1992 May-Jun;12(3):709-17.

[30] Fletcher RH. Carcinoembryonic antigen. *Ann Intern Med.* 1986 Jan;104(1):66-73.

[31] Ishii M. Limitation of clinical usefulness of tumor marker. *Gan. To Kagaku Ryoho.* 1995 Aug;22(9):1139-45.

[32] Motoo Y, Watanabe H, Sawabu N._Sensitivity and specificity of tumor markers in cancer diagnosis. *Nippon Rinsho.* 1996 Jun;54(6):1587-91.

[33] Evans R, Edwards AG, Elwyn G, Watson E, Grol R, Brett J, Austoker J. "It's a maybe test": men's experiences of prostate specific antigen testing in primary care. *Br. J. Gen. Pract.* 2007 Apr;57(537):303-10.

[34] Marteau TM, Kidd J, Michie S, Cook R, Johnston M, Shaw RW. Anxiety, knowledge and satisfaction in women receiving false positive results on routine prenatal screening: a randomized controlled experiment. *J. Psychosom. Obstet. Gynaecol.* 1993 Sep;14(3):185-96.

[35] Stockmann W, Bablok W, Luppa P. Analytical characteristic of Elecsys 2010--a multicentre evaluation. *Wien Klin Wochenschr.* 1998;110 Suppl 3:10-21.

[36] van Ingen HE, Chan DW, Hubl W, Miyachi H, Molina R, Pitzel L, Ruibal A, Rymer JC, Domke I. Analytical and clinical evaluation of an electrochemiluminescence immunoassay for the determination of CA 125. *Clin. Chem.* 1998 Dec;44(12):2530-6.

[37] Hubl W, Chan DW, Van Ingen HE, Miyachi H, Molina R, Filella X, Pitzel L, Ruibal A, Rymer JC, Bagnard G, Domke I. Multicenter evaluation of the Elecsys CA 125 II assay. *Anticancer Res.* 1999 Jul-Aug;19(4A):2727-33.

[38] Stieber P, Molina R, Chan DW, Fritsche HA, Beyrau R, Bonfrer JM, Filella X, Gornet TG, Hoff T, Jager W, van Kamp GJ, Nagel D, Peisker K, Sokoll LJ, Troalen F, Untch M, Domke I. Evaluation of the analytical and clinical characteristic of the Elecsys CA 15-3 immunoassay. *Clin. Chem.* 2001 Dec;47(12):2162-4.

[39] Stern P, Bartos V, Vavrova J, Bezdickova D, Pechova M, Uhrova J, Friedecky B, Sprongl L, Zima T, Palicka V. Comparability of eight immunoassay procedures for the determination of CA 15-3 and related markers. *Clin. Chem. Lab. Med.* 2003 Aug;41(8):1087-94.

[40] Stern P, Friedecky B, Bartos V, Bezdickova D, Vavrova J, Uhrova J, Rozprimova L, Zima T, Palicka V. Comparison of different immunoassays for CA 19-9. *Clin. Chem. Lab. Med.* 2001 Dec;39(12):1278-82.

[41] Passerini R, Riggio D, Salvatici M, Zorzino L, Radice D, Sandri MT. Interchangeability of measurements of CA 19-9 in serum with four frequently used assays: an update. *Clin. Chem. Lab. Med.* 2007;45(1):100-4.

[42] Sanchez-Carbayo M, Espasa A, Chinchilla V, Herrero E, Megias J, Mira A, Soria F. New electrochemiluminescent immunoassay for the determination of CYFRA 21-1: analytical evaluation and clinical diagnostic characteristic in urine samples of patients with bladder cancer. *Clin. Chem.* 1999 Nov;45(11):1944-53.

[43] Filella X, Alcover J, Molina R, Luque P, Corral JM, Auge JM, Coca F. Usefulness of proprostate-specific antigen in the diagnosis of prostate cancer. *Anticancer Res.* 2007 Jan-Feb;27(1B):607-10.

[44] Lein M, Kwiatkowski M, Semjonow A, Luboldt HJ, Hammerer P, Stephan C, Klevecka V, Taymoorian K, Schnorr D, Recker F, Loening SA, Jung K. A multicenter clinical experiment on the use of complexed prostate specific antigen in low prostate specific antigen concentrations. *J. Urol.* 2003 Oct;170(4 Pt 1):1175-9.

[45] Gray MA, Cooke RR, Weinstein P, Nacey JN. Comparability of serum prostate-specific antigen measurement between the Roche Diagnostics Elecsys 2010 and the Abbott Architect i2000. *Ann. Clin. Biochem.* 2004 May;41(Pt 3):207-12.

[46] Vollmer RT, Govindan R, Graziano SL, Gamble G, Garst J, Kelley MJ, Christenson RH. Serum CYFRA 21-1 in advanced stage non-small cell lung cancer: an early measure of response. *Clin. Cancer Res.* 2003 May;9(5):1728-33.

[47] Muley T, Ebert W, Stieber P, Raith H, Holdenrieder S, Nagel D, Furst H, Roth HJ, Luthe H, Blijenberg BG, Gurr E, Uhl W, von Pawel J, Drings P. Technical characteristic and diagnostic utility of the new Elecsys neuron-specific enolase enzyme immunoassay. *Clin. Chem. Lab. Med.* 2003 Jan;41(1):95-103.

[48] Owen WE, Roberts WL. Comparison of five automated serum and whole blood folate assays. *Am. J. Clin. Pathol.* 2003 Jul;120(1):121-6.

[49] Vogeser M, Lorenzl S. Comparison of automated assays for the determination of vitamin B(12) in serum. *Clin. Biochem.* 2007 Aug 11;

[50] Snyder JA, Rogers MW, King MS, Phillips JC, Chapman JF, Hammett-Stabler CA. The impact of hemolysis on Ortho-Clinical Diagnostic's ECi and Roche's elecsys immunoassay systems. *Clin. Chim. Acta.* 2004 Oct;348(1-2):181-7.

[51] Forest JC, Masse J, Lane A. Evaluation of the analytical characteristic of the Boehringer Mannheim Elecsys 2010 immunoanalyzer. *Clin. Biochem.* 1998 Mar;31(2):81-8.

[52] Peetz D, Schweigert R, Jachmann N, Post F, Schinzel H, Lackner KJ. Method comparison of cardiac marker assays on PATHFAST, StratusCS, AxSYM, Immulite 2000, triage, elecsys and cardiac reader. *Clin. Lab.* 2006;52(11-12):605-14.

[53] Hetland O, Dickstein K. Cardiac troponin T by Elecsys system and a rapid ELISA: analytical sensitivity in relation to the TropT (CardiacT) "bedside" test. *Clin. Chem.* 1998 Jun;44(6 Pt 1):1348-50.

[54] Bionda C, Pettazzoni M, Rodriguez-Lafrasse C, Ardail D, Rousson R. Cardiac troponin I and C: analytical comparison and clinical-biological interpretation of three troponins assays. *Ann. Biol .Clin.* (Paris). 2005 Mar-Apr;63(2):185-92.

[55] Heeschen C, Deu A, Langenbrink L, Goldmann BU, Hamm CW. Analytical and diagnostic characteristic of troponin assays in patients suspicious for acute coronary syndromes. *Clin. Biochem.* 2000 Jul;33(5):359-68.

[56] Goldsmith SJ. Radioimmunoassay: review of basic principles. *Semin. Nucl. Med.* 1975 Apr;5(2):125-52.

[57] Schwartz IL. Solomon A. Berson and Rosalyn S. Yalow: a scientific appreciation. *Mt Sinai J. Med.* 1973 May-Jun;40(3):284-94.

[58] Yalow RS. Radioimmunoassay. Practices and pitfalls. *Circ Res.* 1973 May 5;32:Suppl 1:116-28.

[59] Skelley DS, Brown LP, Besch PK. Radioimmunoassay. *Clin. Chem.* 1973 Feb;19(2):146-86.

[60] Parker CW. Radioimmunoassays. *Prog Clin Pathol.* 1972;4(0):103-41.

[61] Berson SA, Yalow RS. General principles of radioimmunoassay. *Clin. Chim. Acta.* 1968 Sep;22(1):51-69.

[62] Irie M. Clinical biochemistry with radioimmunoassay (author's transl). *Seikagaku.* 1979 Jul 25;51(7):505-25.

[63] Pratt JJ, Woldring MG. Radioassay of vitamin B12 and other corrinoids. *Methods Enzymol.* 1982;84:369-406.
[64] Christensen MS. Radioimmunoassay of human parathyroid hormone. A review with special reference to hyperparathyroidism in renal disease. *Dan. Med. Bull.* 1979 Jul;26(4):157-74.
[65] Binkley N. Vitamin D: clinical measurement and use. J. *Musculoskelet Neuronal. Interact.* 2006 Oct-Dec;6(4):338-40
[66] Hao L, Tang Y, Li Z. Progress of the study of determination of folic acid. *Zhonghua Yu Fang Yi Xue Za Zhi.* 1999 May;33(3):177-9.
[67] Kariyone S. Present status of nuclear hematology. *Rinsho Hoshasen.* 1986 Mar;31(3):349-57.
[68] Lee DS, Griffiths BW. Human serum vitamin B12 assay methods--a review. *Clin. Biochem.* 1985 Oct;18(5):261-6.
[69] Mollin DL, Anderson BB, Burman JF. The serum vitamin B12 level: its assay and significance. *Clin. Haematol.* 1976 Oct;5(3):521-46.

Chapter 13

NANOIMAGING AND NANO-VISUALIZATION

INTRODUCTION TO "SMALL OBJECT" IMAGING TECHNOLOGY [1 – 4]

It is accepted that the visual is better than speed. For description of an object, one might use hundreds of words to bring the correct understanding to the listener but the process can be easier with only one picture. In medicine, a picture is very important. First, a picture can be the best tool for teaching and education. There is a classical oriental idiom that "Ten listening is not equal to seeing." Many diseases or disorders have their specific manifestations. Their characteristics require a pictorial presentation to the new physician in order to understand them. Second, the diagnosis in medicine usually requires a picture of a clinical manifestation for making a differential diagnosis. There is a specific biomedical science namely clinical microscopy. This biological science focuses on the diagnosis based on a microscopic picture.

As a tool, the microscope is an important scientific device for visualization of small objects. Combined with advanced photographic technology, the microscopic picture can be produced for usage. In addition to clinical microscopy, there is another biomedical science involving the picture. That science is radiodiagnosis. However, it focuses on large, not small, objects such as bone and tissue. Nevertheless, the applied use of a radiological technique for visualizing small objects can be seen is the case of radiomicroscopy.

In engineering, imaging technology is also very important. The main focus of imaging technology in engineering is the study of material. Microscopy is commonly used in material science. Material analysis is important in construction of a building. Nevertheless, the use of radiological methods for the study of polymers is an important activity of chemical engineering. This radiological investigation, especially for crystallography, is generally applied for real indusexperiment usage. The petrochemical industry is the best example of an industry that makes use of polymer crystallography research.

Table 1. Example of important applied usage of imaging technology in medicine and engineering

In medicine	In engineering
Microbiological diagnosis	Crystal structure detection in chemical engineering
Clinical microscopy diagnosis	Material analysis in civil engineering and electronic engineering

Naturally, an image or picture can be received or perceived by eyes, the light receptor of human beings. The pictorial signal will be transferred to the cerebral cortex for further interpretation [5]. This is the natural process for visualization. Basically, there might be three components for the whole visualization process. First, there must be an "object" which is the thing that poses its own "picture". Second there must be the transferring medium which is light. Light impacts the "object" then it reflects to the eye of human beings. Third, the receptor, which means the photoreceptors in the eye, is required. However, the system can be modified; now there is an electronic sensor that can be the photoreceptor.

However, the visualization process seems to be an individual's perception. To generate a thing that can be similarly visualized and perceived by others is the process of picture or image making. Imaging is the way to support this need. The classical sample imaging technique is painting or drawing, which is actually not similar to the real object. Therefore, a more advanced method such as the photographical one was developed. In photographyl a photo can be generated by applied usage of photosensitive materials on a paper that will catch reflected light from the object. Storrow et al. said that photography required an expedient, portable, adaptable, and relatively simple camera system to take usefulness of fleeting chances for recording visually educational material [6]. They also noted that these prerequisites in an emergency unit were different from those for traditional medical photography, for which relatively plentiful time and advanced equipment were routinely available [6]. They also mentioned that medical photography departments had to review photographic equipment, paying attention to speed, ease of use, and quality of output as well as basic simple ways such as film choice, lighting, close-up photography, standardization, copy work, and radiographs [6]. With the latest advances, now photos can be directly captured into an electronic format, which makes them reproducible [7]. The image processing can be performed. This new advent is very useful in the scientific world.

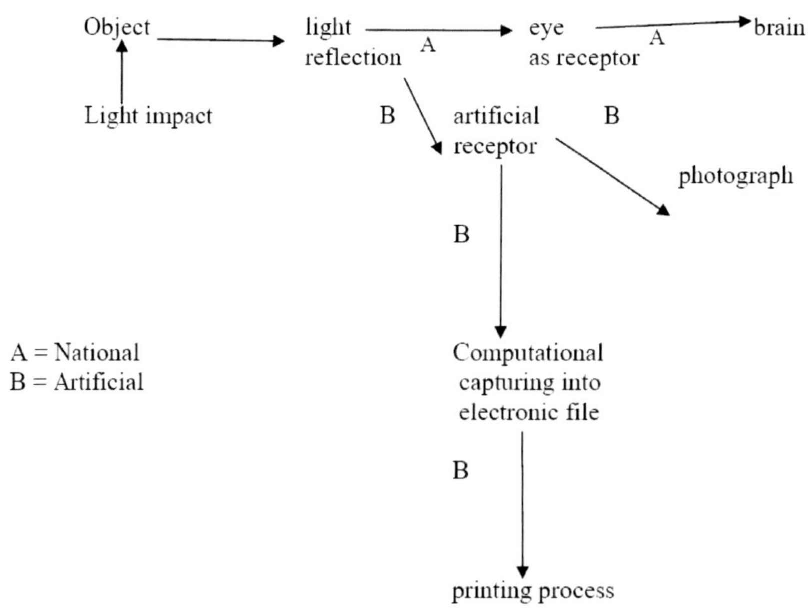

Figure 1. Diagram showing visualization and imaging.

Considering a picture, there are many properties that can be used for image processing. Color, shape, control, and texture are the main properties. Based on a picture's property analysis, discrimination between two or many similar pictures can make use of analytical intelligence for discrimination of problematic similar pictures. A good sample of this applied usage is the usage in an automated urinalysis analyzer, namely IQ-200 in medicine [8].

MICROSCOPE AS A TOOL FOR VISUALIZATION OF SMALL OBJECTS

It is accepted that human eyes have limited ability for visualization. Very small objects below a millimeter level cannot be naturally visualized. Scientists have a lot of interest in very small objects and tried to invent a new tool to help see the micro-objects. The of microscope intervention was many centuries ago. Developed from a natural light microscope, many advanced microscopes have been invented and are in use now. The main direction of microscope development is increasing the ability to visualize smaller objects. With this aim, new light sources, including ultraviolet light, X-ray and electron beam, are applied. With these new advents, present microscopes can read the nanolevel structure. The details on important types of microscopes that can be used for nanolevel structure visualization will be further described.

1. Scanning Electron Microscope (SEM) [9 - 18]

SEM is a subtype of electron microscope. The light source is the electron beam.

SEM is useful for studying the surface of an object. It is commonly used for visualization of surfaces of materials at nanolevel. The methods for handling the specimen, stabilization, dehydration and drying, rendering the surface conductive, and exposing the surface of interest are described for a wide multiple appearance of experimental animals [19]. Katterwe et al said that problems emanating from restricted depth of focus in light-microscopical ways were eliminated by utilizing the scanning electron microscope [20]. Various attempts have been made to make the scanning electron microscope suitable for comparison microscopy: videotape, image storage tube, split field scan [20]. In 1980 another way of performing scanning electron comparison microscopy was contrived: the coupling of two tools. Both tools are equipped with many electron detection systems, which produce special contrast effects [20]. A recent development in environmental SEM (ESEM) allows the observation of nano-objects suspended in a liquid phase, with a few nanometers resolution and a good signal to noise ratio [21]. The idea behind this way is simply to perform STEM-in-SEM, that is SEM in transmission mode, in an environmental SEM [21]. This new way extends the scope of SEM as a high-resolution microscope, a relatively cheap and generalized available imaging tool [21]. Some interesting reports in using SEM in nanobiotechnology are presented in Table 1.

Table 1. Reports in using SEM in nanobiotechnology

Authors	Details
Shibli et al. [22]	In the present work, a new electrochemical sensor was developed by electroplating nano nickel oxide reinforced nickel on graphite substrate [22]. In this work, the morphology of the working electrode surface was studied by using a SEM [22]. Shibli et al. said that the modified electrode revealed good sensing characteristic with a response time as low as 8s during sensing and estimation of acetylcholine [22].
Wang et al. [23]	Zein was studied as a medicine-eluting coating film composed of zein microshperical objects for cardiovascular devices [23]. SEM observation indicated that the zein film was made of microshperical objects in diameter from nano- to micrometer, which could be controlled [23]. Wang et al. suggested that zein film could be used directly as a new type of coating material for its better biocompatibility with human umbilical veins endothelial cells [23].
Fujihara et al. [24]	In this study, a new type of guided bone regeneration (GBR) membranes were fabricated by polycaprolactone (PCL)/CaCO3 composite nano-fibers with two different PCL to calcium carbonate (CaCO3) ratios [24]. In this experiment, two different GBR membranes were treated, i.e., GBR membrane (A)=PCL:CaCO3=75:25 wt%+PCL, GBR membrane (B)=PCL:CaCO3=25:75 wt%+PCL [24]. SEM observation revealed that no significant difference in osteoblast attachment manner was seen on GBR membrane (A) and (B) [24]. Fujihara et al. concluded that because of good cell attachment manner, there was a potential to utilize PCL/CaCO3 composite nano-fibers to GBR membranes [24].
Zhang et al. [25]	In this work, completely sulfonated polyaniline nano-particles, nano-fibrils and nano-networks have been getd for the first time by electrochemical homopolymerization of orthanilic acid using a three-step electrochemical deposition procedure in a mixed solvent of acetonitrile and hydro [25]. The details of the poly(orthanilic acid) nano-structure were examined with a field emission SEM [25].

2. Transmission Electron Microscope (TEM)

TEM is another subtype of electron microscope. The light source is the electron beam similar to that of SEM but the TEM is based on transmission light. It is commonly used for study of the cut-off surface of an object. However, a big problem of using TEM is the fact that hydro-containing biological material cannot withstand the vacuum of the transmission electron microscope [26]. The classical solution is to dehydrate chemically fixed biological samples and then embed them in resin [26]. Before TEM analysis, insulating specimen needs to become coated with a charge-draining layer [27]. The introduction of freeze-substitution based on dehydration of unfixed cryo-immobilized samples at low temperature can substantially reduce aggregation [26]. Hoche et al noted that image quality could be amplified in high-resolution imaging and sensitivity could be significantly increased in all nano-beam related ways including elemental analytics, convergent-beam and nano-beam electron diffraction, and spectral imaging [27]. Some interesting reports in using TEM in nanobiotechnology are presented in Table 2.

Table 2. Reports in using TEM in nanobiotechnology

Authors	Details
Haung et al. [28]	In this study, the tubule formation process from amorphous carbon nanowires under high-bias-caused Joule heating was observed in real time in a high-resolution transmission electron microscope [28]. Haung et al. noted that crystallization of the amorphous carbon nanowires occurred in two distinct ways as a) the formation of tubular graphitic basal planes parallel to the nanowire axis on the surface and b) the formation of nano-onions in the interior of the nanowire [28].
Jurgens et al. [29]	In order to visualise protein molecules organised in films on surfaces of nano-sized gold-particles, an electron-microscopic approach based on the scattering absorption contrast of the specimen was applied by Jurgens et al. [29]. In this work, an electron-microscopic approach based on the scattering absorption contrast of the specimen should be applied in order to visualize protein molecules organized in films on surfaces of nano-sized gold-particles [29].
Yamasaki et al. [30]	Yamasaki et al. reported the first experiments of selected area nano-diffraction from semiconductor interfaces using a spherical aberration corrected TEM [30]. Yamasaki et al. succeeded in obtaining sharp diffraction patterns from well-defined areas on both sides of the interface between a Ti-Si-Ge alloy and Si0.54Ge0.46 [30]. In this work, the errors in the area selection by an SAD aperture were reduced to less than 2 nm by the Cs-correction [30].
Akita et al. [31]	In this work, Au/CeO(2) and Ir/CeO(2) catalysts were observed by a transmission electron microscope in order to investigate the nano-structures of Au and Ir particles and CeO(2) grains and the interface structure between metallic particles and CeO(2) [31].
Kamino et al. [32]	In this work, interaction between multi-walled carbon nanotubes (MWNTs) and deposited gold nano-particles has been dynamically observed in a 200 kV TEM using a specimen heating holder [32]. It was found that some gold nano-particles drilled through the wall of the MWNT and entered the capillary space of the MWNTs [32].

3. Scanning Transmission Electron Microscope (STEM)

This is a combination between SEM and TEM. In the STEM, an electron beam of a few angstroms diameter is raster scanned over a thin sample and the scattered electrons are sequentially determined for each sample element irradiated [33]. With STEM, the mass, the elemental composition and the structure of a protein can be assessed [33]. STEM is accepted for its good efficiency of the different elastic and inelastic signals resulting from the interaction of the high energy primary electrons with the specimen [34]. As previously mentioned for ESEM, a new imaging mode called "wet-STEM" was also developed with a similar objective [21]. Some interesting reports in using STEM in nanobiotechnology are presented in Table 3.

Table 3. Reports in using STEM in nanobiotechnology

Authors	Details
Dong et al. [35]	This paper reported the analysis of plasma-sprayed HA/Ti-6Al-4V composites by TEM and STEM to elucidate the intricate nature of the materials following plasma spray processing and in vitro evaluation [35]. After immersion in simulated body fluid for 2 weeks and 10 weeks, TEM and STEM warranted that the interfaces inside the coating maintained good microstructural integrity [35].
Shimojo et al. [36]	Shimojo et al. reported on an applied usage of STEM to nanometre-sized fabrication by means of electron beam-induced deposition. In this work, electron beam-induced deposition was carried out using a STEM with a field emission gun to fabricate nanometre-sized structures.

4. Atomic Force Microscope (AFM) [37 - 38]

AFM is a commonly used tool in nanodiagnosis. The light source is laser. AFM can be used for study of height or altitude pattern of the surface of the studied object at nanolevel. The measurement is the determination of atomic force on each atom. Basically, the light source, laser, is focused on a cantilever which has a tip. The tip atoms are the direct part of the AFM that contact with surface atoms of the determined object. The counteraction form from the surface atom was transferred through the cantilever and detected via the reflection of laser pattern that is further caught by photodetector. At present, AFM images have progressed from relatively crude representations of specimen topography to the nanometer scale representations of native hydrated surfaces [39]. The potential of the AFM, which does not require a conductive sample, can be shown with molecular resolution images of a nonconducting organic monolayer and an amino acid crystal [38]. Applied usage of this new microscope to technology is demonstrated with images of an optical disk stamper, a diffraction grating, a thin-film magnetic recording head, and a diamond cutting tool [38]. The usefulness of the AFM for biologists is that AFM can visualize non-conductive materials in a non-vacuous, such as air or liquid, environment [40]. Despite many uses, AFM of biological specimens at room temperature is still severely limited by at least two factors - the softness and the thermal motion of flexible multi-domain/subunit molecules [41]. Luckily, both problems can be overcome by imaging biological structures at cryogenic temperatures [41].

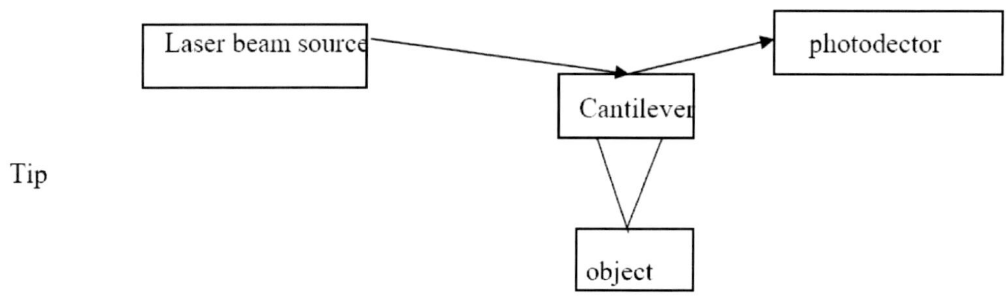

Figure 2. Diagram showing the analytical process of atomic force microscoper.

As previously mentioned, AFM is useful for diagnosis. Many diseases can be diagnosed via AFM. For example, the change of red blood cell membrane pattern observed by AFM can be useful for diagnosis of malarial infection [42 – 43]. el-Shoura reported that anemia was common in patients with AFM which was usually associated with structural alterations of non-parasitized immature and mature red blood cells [43]. The diagnosis of cartilage surface can also be useful for diagnosis of osteoarthritis.

It can be seen that the altitude level on the surface of cartilage in healthy subjects has the difference in height less than that of osteoarthritic subjects [44 - 45]. The other examples are determination of myosin in motor protein defect [46 - 48]. In addition to visualization, AFM can also be applied as a tool for rearrangement of atoms on a material surface.

5. Scanning Near-Field Optical Microscopy (SNOM)

Scanning a point light source in close proximity over a sample and recording the scattered or transmitted light intensity point-by-point allows one to record optical images with a resolution not limited by diffraction [49]. SNOM is based on this principle. With SNOM, a picture of DNA and chromosome can be easily derived [50]. Yoshino et al mentioned that fluorescence-stained lambda-phage DNA analysis by SNOM/AFM demonstrated the possibility of nanometer-scale imaging for a new way termed nano-fluorescence in situ hybridization (nano-FISH) [50].

Visual Image Analysis [51 – 55]

As previously mentioned, visual image analysis is a useful technique in visual image processing. It can be the basis for development of a database via a search engine. The basic concept that should be known is the meaning of image in visual image processing field. Basically, the image can be divided into physical image and digital image. Visual image is the product of optical process according to light or electromagnetic spectrum at different frequencies (X-ray, Infextremely rared, sound, etc.) while digital image is a two-dimensional function of light intensity. Digital image is used for visual image processing. The process starts from image acquisition, which is the step to change the captured continuous three-dimensional image into a two-dimensional image. The next step is image sampling to obtain the function of picture element for pel or pixel. The final step is quantization of color tone for each picture element.

Based on the described steps of the image digitalization process, a digital image can be derived. To set up a system for image retrieval is the next important step in making a search engine for digital images. The discrimination between different images can be accepted based on many properties including uniqueness, completeness, sensitivity, distraction from detail, and geometrical transformation invariance (including transsite, scaling, rotation and reflection). The discrimination functions can be shape descriptor, color descriptor, texture descriptor or special relation as once previously mentioned. Based on described descriptions, the database of digital images can be set up and image retrieval can be easily performed. The way can be useful for setting up the database and system for discrimination of small nanoobjects, which are very hard to discrete by a simple optical discrimination way.

Quantum Dot Imaging [56 – 63]

Of many nanoimaging ways, quantum dot is a new concern. Quantum dot is a crystalline fluorophores with CdSe semiconductor core and ZnS shell. It has hydrophobic crystals with unique spectral properties (broad absorption and narrow emission). The wavelength of quantum dot depends on size. Since starting pointal synthesized quantum dot is metal and not hydro soluble, various methods for making them hydro-soluble have been developed. Those methods include derivatizing surface with mercaptoacetic acid, encapsulating in phospholipid micelles, derivatizing surface with DHLA and coating them with amine-modified. In addition, conjugating quantum dot to biomolecules is also feasible. Quantum dot is photostable and lasts longer than a fluorescence probe. Quantum dot also allows the scientist to see the physiological change within the labeled cell (due to its long lasting property). The usefulness of quantum dot include a) specific tagging of cells and tissues, b) useful for long-term imaging, c) no cytotoxicity observed in vitro/in vivo, d) useful for multi-color multiplexing, and e) suitable for dynamic imaging of subcellular structures. Problems with its use include a) colloidal polymer-coated quantum dots can aggregate irreversibly, b) quantum dots are bulkier than many organic fluorophores and c) the size of modified quantum dot makes it not feasible to pass into the cell. Due to the increased concern about the probable in vivo side effect of quantum dot, most quantum dot are available in modified coated form. More details on quantum dot can be seen in another chapter on quantum medicine in this book.

Nanomagnetic Imaging

Nanomagnetic imaging is another advent in nanomedicine. Superparamagnetic phenomenon is the basis for nanomagnetic imaging. Changing of magnetic status due to temperature is the main observation in this phenomenon. With advent in nanomaterial, nanomaterials with supramagnetic property can be synthesized [64 - 65]. These new materials can be used for nanoimaging. Superparamagnetic iron oxide nanocrystals are extensively studied as excellent magnetic resonance imaging (MRI) probes for various cell trafficking, gene expression, and cancer diagnosis [66]. For example, in 2005, Huh et al. reported an in vivo diagnosis of cancer, utilizing a well-defined magnetic nanocrystal probe system with multiple capabilities, such as small size, strong magnetism, high biocompatibility, and the possession of active functionality for desired receptors [66]. In 2007, Lee et al. used artificial engineering approaches to develop innovative magnetic nanoprobes, through a process that involved the systematic evaluation of the magnetic spin, size and type of spinel metal ferrites. Lee et al reported that those magnetism-engineered iron oxide nanoprobes, when conjugated with antibodies, revealed amplified magnetic resonance imaging sensitivity for the detection of cancer markers compared with available probes [67].

In addition to imaging, the nanomagnetic particle can also be applied for treatment. For example, Fan et al. recently reported preparation and characterization of magnetic nanoparticles with radiofrequency-induced hyperthermia for cancer treatment [68]. In another study by Giri et al., ultrafine particles of substituted ferrite Co(1-a)Zn(a)Fe2O4 and substituted yttrium-iron garnet Y3Fe(5-x)Al(x)O12 were treated through microwave refluxing and citrate-gel route respectively for further hyperthermia applied usages [69].

REFERENCES

[1] Uversky VN, Kabanov AV, Lyubchenko YL.Nanotools for megaproblems: probing protein misfolding diseases using nanomedicine modus operandi. *J. Proteome. Res.* 2006 Oct;5(10):2505-22.

[2] Lyubchenko YL, Sherman S, Shlyakhtenko LS, Uversky VN. Nanoimaging for protein misfolding and related diseases. *J. Cell Biochem.* 2006 Sep 1;99(1):52-70.

[3] Fankhauser F, Niederer PF, Kwasniewska S, van der Zypen E. Supernormal vision, high-resolution retinal imaging, multiphoton imaging and nanosurgery of the cornea--a review. *Technol. Health Care.* 2004;12(6):443-53.

[4] Soon L, Braet F, Condeelis J. Moving in the right direction-nanoimaging in cancer cell motility and metastasis. *Microsc. Res. Tech.* 2007 Mar;70(3):252-7.

[5] Hirose K. Imaging the functions of central neurons. *Nihon Shinkei Seishin Yakurigaku Zasshi.* 2006 Jun;26(3):125-9.

[6] Storrow AB, Stack LB, Petersen P. An approach to emergency department photography. *Acad. Emerg. Med.* 1994 Sep-Oct;1(5):454-62.

[7] Wiwanitkit V. Digital camera and picture capturing computer program for microscopic picture creation : a millenium in clinical microscopy. *Taksin Med. J.* 1999; 17(2): 88- 92.

[8] Alves L, Ballester F, Camps J, Joven J. Preliminary evaluation of the Iris IQ 200 automated urine analyser. *Clin. Chem. Lab. Med.* 2005;43(9):967-70.

[9] Hollenberg MJ, Erickson AM. The scanning electron microscope: potential usefulness to biologists. A review. *J. Histochem. Cytochem.* 1973 Feb;21(2):109-30.

[10] Zinn KM, Smith RS. Specialized ways in electron microscopy. *Int. Ophthalmol. Clin.* 1973 Fall;13(3):33-49.

[11] Hayes TJ. Potentials and prospects in scanning electron microscopy. *J. Microsc.* 1974 Mar;100(2):133-42.

[12] Mironov AA, Mironov VA, Rekhter MD. Simplified methods of preparing biological objects for scanning electron microscopy. *Arkh. Anat. Gistol. Embriol.* 1983 Nov;85(11):85-9.

[13] Muir MD. The Scanning Electron Microscope Symposium April 23-27, 1973, Chicago, Illinois. *Microsc. Acta.* 1973 Sep;75(1):62-8.

[14] Turner RH, Smith B. A simple way for examining fresh, frozen, biological specimens in the scanning electron microscope. *J. Microsc.* 1974 Nov;102(2):209-14.

[15] Messier PM. Scanning electron microscopy. *Union. Med. Can.* 1974 Apr;103(4):727-31.

[16] Nowell JA, Pawley JB. Preparation of experimental animal tissue for SEM. *Scan Electron Microsc.* 1980;(Pt 2):1-19.
[17] Pomeranz Y. Scanning electron microscopy in food science and technology. *Adv. Food Res.* 1976;22:205-307.
[18] Carr KE. Applied usages of scanning electron microscopy in biology. *Int. Rev. Cytol.* 1971;30:183-255.
[19] Heinzmann U. Refined preparation of biological specimens for the scanning-electron-microscope. *Microsc. Acta.* 1974 Sep;76(2):145-6.
[20] Katterwe H, Goebel R, Grooss KD. The comparison scanning electron microscope within the field of forensic science. *Scan Electron. Microsc.* 1982;(Pt 2):499-504.
[21] Bogner A, Jouneau PH, Thollet G, Basset D, Gauthier C. A history of scanning electron microscopy developments: towards "wet-STEM" imaging. *Micron.* 2007;38(4):390-401.
[22] Shibli SM, Beenakumari KS, Suma ND. Nano nickel oxide/nickel integrated nickel composite coating for sensing and estimation of acetylcholine. *Biosens Bioelectron.* 2006 Dec 15;22(5):633-8.
[23] Wang HJ, Lin ZX, Liu XM, Sheng SY, Wang JY. Heparin-loaded zein microsphere film and hemocompatibility. *J. Control. Release.* 2005 Jun 20;105(1-2):120-31.
[24] Fujihara K, Kotaki M, Ramakrishna S. Guided bone regeneration membrane made of polycaprolactone/calcium carbonate composite nano-fibers. *Biomaterials.* 2005 Jul;26(19):4139-47.
[25] Zhang L, Jiang X, Niu L, Dong S. Syntheses of completely sulfonated polyaniline nano-networks and its applied usage to the direct electrochemistry of cytochrome c. *Biosens Bioelectron.* 2006 Jan 15;21(7):1107-15.
[26] Eltsov M, Zuber B. Transmission electron microscopy of the bacterial nucleoid. *J. Struct. Biol.* 2006 Nov;156(2):246-54.
[27] Hoche T, Gerlach JW, Petsch T. Static-charging mitigation and contamination avoidance by selective carbon coating of TEM samples. *Ultramicroscopy.* 2006 Oct-Nov;106(11-12):981-5.
[28] Huang JY, Chen S, Ren ZF, Chen G, Dresselhaus MS. Real-time observation of tubule formation from amorphous carbon nanowires under high-bias Joule heating. *Nano Lett.* 2006 Aug;6(8):1699-705.
[29] Jurgens L, Nichtl A, Werner U. Electron density imaging of protein films on gold-particle surfaces with transmission electron microscopy. *Cytometry.* 1999 Oct. 1;37(2):87-92.
[30] Yamasaki J, Sawada H, Tanaka N. First experiments of selected area nano-diffraction from semiconductor interfaces using a spherical aberration corrected *TEM. J. Electron. Microsc.* (Tokyo). 2005 Apr;54(2):123-6.
[31] Akita T, Okumura M, Tanaka K, Kohyama M, Tsubota S, Haruta M. TEM observations of Au and Ir particles supported on CeO2. *J. Electron. Microsc.* (Tokyo). 2005;54 Suppl 1:i81-i85.
[32] Kamino T, Yaguchi T, Konno M, Hashimoto T. In situ high temperature TEM observation of interaction between multi-walled carbon nanotube and in situ deposited gold nano-particles. *J. Electron. Microsc.* (Tokyo). 2005 Oct;54(5):461-5.
[33] Muller SA, Engel A. Structure and mass analysis by scanning transmission electron microscopy. *Micron.* 2001 Jan;32(1):21-31.

[34] Colliex C, Mory C. Scanning transmission electron microscopy of biological structures. *Biol. Cell.* 1994;80(2-3):175-80.
[35] Dong ZL, Khor KA, Quek CH, White TJ, Cheang P. TEM and STEM analysis on heat-treated and in vitro plasma-sprayed hydroxyapatite/Ti-6Al-4V composite coatings. *Biomaterials.* 2003 Jan;24(1):97-105.
[36] Shimojo M, Mitsuishi K, Tanaka M, Han M, Furuya K. Applied usage of transmission electron microscopes to nanometre-sized fabrication by means of electron beam-induced deposition. *J. Microsc.* 2004 Apr;214(Pt 1):76-9.
[37] Ikai A. Biomacromolecules studied by atomic force microscopy. *Nippon Rinsho.* 1993 Oct;51(10):2761-8.
[38] Hansma PK, Elings VB, Marti O, Bracker CE. Scanning tunneling microscopy and atomic force microscopy: applied usage to biology and technology. *Science.* 1988 Oct 14;242(4876):209-16.
[39] Firtel M, Beveridge TJ. Scanning probe microscopy in microbiology. *Micron.* 1995;26(4):347-62.
[40] Ushiki T, Hitomi J, Ogura S, Umemoto T, Shigeno M. Atomiic force microscopy in histology and cytology. *Arch. Histol. Cytol.* 1996 Dec;59(5):421-31.
[41] Shao Z, Zhang Y. Biological cryo atomic force microscopy: a brief review. *pUltramicroscopy.* 1996 Dec;66(3-4):141-52.
[42] Aikawa M. Studies on falciparum malaria with atomic-force and surface-potential microscopes. *Ann .Trop Med. Parasito*l. 1997 Oct;91(7):689-92
[43] el-Shoura SM. Falciparum malaria in naturally infected human patients: ultrastructural alterations of non-parasitized red blood cells during anaemia. *Appl. Parasitol.* 1993 Sep;34(3):173-9.
[44] Hassenkam T, Jorgensen HL, Lauritzen JB. Mapping the imprint of bone remodeling by atomic force microscopy. *Anat. Rec. A Discov. Mol. Cell Evol. Biol.* 2006 Oct;288(10):1087-94.
[45] Gonzalez S, Fragoso-Soriano RJ, Kouri JB. Chondrocytes interconnecting tracks and cytoplasmic projections observed within the superficial zone of normal human articular cartilage-A transmission electron microscopy, atomic force microscopy, and two-photon excitation microscopy studies. *pMicrosc Res. Tech.* 2007 Sep 12;
[46] Taniguchi M, Matsumoto O, Suzuki S, Nishino Y, Okuda A, Taga T, Yamane T. MgATP-induced conformational changes in a single myosin molecule observed by atomic force microscopy: periodicity of substructures in myosin rods. *Scanning.* 2003 Sep-Oct;25(5):223-9.
[47] Zhang J, Wang YL, Gu L, Pan J. Atomic force microscopy of actin. *Sheng Wu Hua Xue Yu Sheng Wu Wu Li Xue Bao* (Shanghai). 2003 Jun;35(6):489-94.
[48] Sheng S, Gao Y, Khromov AS, Somlyo AV, Somlyo AP, Shao Z. Cryo-atomic force microscopy of unphosphorylated and thiophosphorylated single smooth muscle myosin molecules. *J. Biol. Chem.* 2003 Oct 10;278(41):39892-6.
[49] Meixner AJ, Kneppe H.Scanning near-field optical microscopy in cell biology and microbiology. *Cell Mol. Biol.* (Noisy-le-grand). 1998 Jul;44(5):673-88.
[50] Yoshino T, Sugiyama S, Hagiwara S, Fukushi D, Shichiri M, Nakao H, Kim JM, Hirose T, Muramatsu H, Ohtani T. Nano-scale imaging of chromosomes and DNA by scanning near-field optical/atomic force microscopy. *Ultramicroscopy.* 2003 Oct-Nov;97(1-4):81-7.

[51] Kunt M. Image coding. *Scanning Microsc. Suppl.* 1988;2:1-29.

[52] Burr DC, Morrone MC. The role of features in structuring visual images. *Ciba Found Symp.* 1994;184:129-41.

[53] Wang D. The time dimension for scene analysis. *IEEE Trans Neural Netw.* 2005 Nov;16(6):1401-26.

[54] Victor JD. Analyzing receptive fields, classification images and functional images: challenges with chances for synergy. *Nat. Neurosci.* 2005 Dec;8(12):1651-6.

[55] Berlage T. Analyzing and mining image databases. *Medicine Discov. Today.* 2005 Jun 1;10(11):795-802.

[56] Huo Q. A perspective on bioconjugated nanoparticles and quantum dots. *Colloids Surf. B Biointerfaces.* 2007 Sep 1;59(1):1-10.

[57] Pathak S, Davidson MC, Silva GA. Characterization of the functional binding properties of antibody conjugated quantum dots. *Nano Lett.* 2007 Jul;7(7):1839-45.

[58] Chan WC, Nie S. Quantum dot bioconjugates for ultrasensitive nonisotopic detection. *Science.* 1998 Sep 25;281(5385):2016-8.

[59] Goldman ER, Balighian ED, Mattoussi H, Kuno MK, Mauro JM, Tran PT, Anderson GP. Avidin: a natural bridge for quantum dot-antibody conjugates. *J. Am. Chem. Soc.* 2002 Jun 5;124(22):6378-82.

[60] Tokumasu F, Dvorak J. Development and applied usage of quantum dots for immunocytochemistry of human erythrocytes. *J. Microsc.* 2003 Sep;211(Pt 3):256-61.

[61] de Farias PM, Santos BS, de Menezes FD, de Carvalho Ferreira R, Barjas-Castro ML, Castro V, Lima PR, Fontes A, Cesar CL. Investigation of red blood cell antigens with highly fluorescent and stable semiconductor quantum dots. *J. Biomed. Opt.* 2005 Jul-Aug;10(4):44023.

[62] de Farias PM, Santos BS, de Menezes FD, Ferreira Rde C, Barjas-Castro Mde L, Castro V, Lima PR, Fontes A, Cesar CL. Core-shell CdS/Cd(OH)2 quantum dots: synthesis and bioconjugation to target red cells antigens. *J. Microsc.* 2005 Sep;219(Pt 3):103-8.

[63] Akhtar RS, Latham CB, Siniscalco D, Fuccio C, Roth KA. Immunohistochemical detection with quantum dots. *Methods Mol. Biol.* 2007;374:11-28.

[64] Zhang J, Post M, Veres T, Jakubek ZJ, Guan J, Wang D, Normandin F, Deslandes Y, Simard B. Laser-assisted synthesis of superparamagnetic Fe@Au core-shell nanoparticles. *J. Phys. Chem. B.* 2006 Apr 13;110(14):7122-8.

[65] Huh YM, Jun YW, Song HT, Kim S, Choi JS, Lee JH, Yoon S, Kim KS, Shin JS, Suh JS, Cheon J. Biocompatible and biodegradable polymer nanofibers displaying superparamagnetic properties. *Chemphyschem.* 2005 Aug 12;6(8):1461-5.

[66] Huh YM, Jun YW, Song HT, Kim S, Choi JS, Lee JH, Yoon S, Kim KS, Shin JS, Suh JS, Cheon J. In vivo magnetic resonance detection of cancer by using multifunctional magnetic nanocrystals. *J. Am. Chem. Soc.* 2005 Sep 7;127(35):12387-91.

[67] Lee JH, Huh YM, Jun YW, Seo JW, Jang JT, Song HT, Kim S, Cho EJ, Yoon HG, Suh JS, Cheon J. Artificially engineered magnetic nanoparticles for ultra-sensitive molecular imaging. *Nat Med.* 2007 Jan;13(1):95-9.

[68] Giri J, Ray A, Dasgupta S, Datta D, Bahadur D. Investigation on Tc tuned nano particles of magnetic oxides for hyperthermia applied usages. *Biomed. Mater. Eng.* 2003;13(4):387-99.

[69] Fan X, Zhang D, Zheng J, Gu N, Ding A, Jia X, Qing H, Jin L, Wan M, Li Q. Preparation and characterization of magnetic nano-particles with radiofrequency-induced hyperthermia for cancer treatment. *Sheng Wu Yi Xue Gong Cheng Xue Za Zhi*. 2006 Aug;23(4):809-13.

Chapter 14

SOCIAL ASPECT OF NANOTECHNOLOGY

EMERGING OF NANOPRODUCT [1 – 8]

Nanotechnology is the new facet of science that has led to major developments in the world. Due to the marvelous properties of a nanoobject, a number of applied usages of nanotechnology can be seen at present. Nanotechnology can be applied in medicine, engineering as well as agriculture. Many nanoproducts have been launched into the market in the past few years.

Table 1. Examples of applied usages of nanotechnology at present

Aspect	Details
Medicine	Nanomedicine is applied nanotechnology for diagnosis, treatment and prevention of diseases [9 – 14].
Engineering	There are many advents in nanoengineering such as nanocomputing, nanoelectronics and nanomaterials [15 – 18].
Agriculture	Nanobiotechnology can be applied for production, storing and processing in agriculture [20].

Second forms of nanoproducts can be seen in the present day. All human basic necessities have been impacted by nanoproducts.

A. Food

Food is important for all human beings; we cannot survive without food ingestion. Applied usages of nanotechnology in agriculture can be seen as previously described. There is a need to design and develop equipment that will be able to generate precise and uniform micro or nano capsules in large quantities for indusexperiment applied usages [21]. The reported food vehicles for delivery of encapsulated probiotic bacteria are yoghurt, cheese, ice cream and mayonnaise [20]. Some interesting reports on the applied usage of nanotechnology in food science are presented in Table 2.

Table 2. Reports on applied usage of nanotechnology in food science

Authors	Details
Broberg et al. [21]	A methodology for NMR analysis of low nanomole summations of oligosaccharides fractionated by analytical HPAEC was presented in this work [21]. In this research, arabinoxylan derived oligosaccharides purified by HPAEC-PAD on an analytical column, by single injections, were analyzed with nano-probe NMR and MALDI-TOF MS to bring full structural assignment [21].
Pfenninger et al. [22]	In this work, a complex mixture of isomeric neutral oligosaccharides from pooled human milk was analyzed by nano-electrospray ionization in a quadrupole ion trap mass spectrometer in the negative ion mode [22].
von Bultzingslowen et al. [23]	Sol-gel based optical carbon dioxide sensor employing couple luminophore referencing for applied usage in food packaging technology was reported in this work [23].
Cataldi et al. [24]	The applied usage of activated pulsed amperometric detection for the determination of orotic acid in real samples at a gold working electrode in alkaline solutions, in combination with anion-exchange chromatography, was reported in this work [24].

B. Clothes

Clothes are necessary for all human beings. Now many applied nanotechnology-based production processes are in use in the textile industry. Nanoparticle-coated clothes have been launched and commonly sold. These products are deemed superior in quality to the classical products. The antibacterial property of titanium and silver nanoclothes are good examples [25]. Biodegradable textile products are another important example [26]. At present, nanostructured biodegradable poly(L-lactide-co-epsilon-caprolactone) fabrics can be treated by electrospinning [26].

C. Housing

Everyone should have a house or place to live. Advents in nanotechnology have brought much new knowledge to use in contruction. A major goal of material science is to produce hierarchical materials that are ordered on all length scales, including the nanolevel [27]. Many nanomaterials are produced for novel building.

D. Medicine

As previously mentioned nanopharmacology is a main and rapidly growing applied usage of nanotechnology [28 - 36]. Many new medicines are developed based are nanotechnology aiming at the most effective therapy.

In addition to the four main necessary objects, there are also other nanoproducts in daily use including supplementary nonessential items such as cosmetics and toys. Therefore, we can say that nanoproducts are around us and will become an important part of our daily life soon. It can be a new item in our society. The social aspect of nanotechnology should be a concern. The important social aspects of nanotechnology will be a focus in this chapter.

NANOPRODUCT ECONOMICS AT PRESENT

As already mentioned there are thousands of nanoproducts in the market at present. These products become a new focus for many companies. The investment in nanoproducts has developed rapidly. Indeed, revolutionary and synergistic advances at the interfaces between previously separated fields of science, engineering and areas of relevance are ready to create nano-bio-info-cogno transforming tools [37]. Spohrer and Engelbart specifically explored applied usages of converging technologies for enhancing human characteristic, that is, enhancing human ability to get goals both individually and collectively [38]. Spohrer and Engelbart proposed that these applied usages held the promise to make people healthier, wealthier, and wiser as well as to make businesses more responsive, resilient, and adaptive [38]. Nanoproduct economics can considered the new phase of the world economy [39 - 40]. The economic equality will be the next challenging problem for the world. The disequilibrium of nanotechnology distribution between developed and producing countries should be solved [41].

ENVIRONMENTAL IMPORTS OF NANOTECHNOLOGY

Because nanotechnology is something new and we just see mainly its usefulness, monitoring of its problems or side effects are necessary. There are many stories in scientific history on unpredictable side effects of a new thing after long-term use such as the congenital defect due to the use of thalidomide. How to prevent a similar situation which might occur due to the use of a nanoproduct should be discussed. A nanotoxicology study is necessary. Although there are reports on the safety of nanoparticles, there might be some underreported cases. In addition, there is no long-term prospective study in this field. Indeed the particles at nanolevel can be seen in nature. The nanovirus is a good example and the nanovirus can be a germ. Many small chemical particles such as benzene have proven to be a carcinogen. For the new synthesized nanomaterials, a closed monitoring is necessary. A relative risk assessment is presented for the industrial fabrication of many nanomaterials [42]. The physical-chemical properties and quantities of the inventoried materials can be used to assess relative risk based on factors such as volatility, carcinogenicity, flammability, toxicity, and persistence [42]. Safety of nanotechnology is the aim of nanoscience at present [43 – 44].

A. Carcinogenesis and Mutagenesis

Carcinogenesis and mutagenesis of nanoparticles is a focus at present because there is no proof that new synthesized nanoparticles are safe and will not be a carcinogen or mutagen. There are many reports indicating that many new synthesized nanoparticles can enter into a human cell. A long-term follow-up on the side effect is necessary. Indeed, there are many nanomaterials that have to be under strict control. The synthesized nanopolymers of hydrocarbon is the best example. Hydrocarbon's high possibility for carcinogenesis and mutagenesis has been proven. Some hydrocarbon polymers are banned. The hydrocarbon-based nanopolymers should be well controlled. An interesting nanopolymer is polyethylimene (PEI) [45 – 46]. PEI is a particle that can be used for gene transferring [47 – 52]. PEI-based nanomaterial is of concern. The main question is whether there is a possibility that the PEI-based nanomaterial of the mutagenic probe could be absorbed into the body. This is a very big question. In addition, since the nanotechnology can be applied to a molecular biology method, a genetically modified object (GMO) as a result can be expected. The GMO is a commonly discussed topic in environmental science [53 – 55]. A big problem is if those GMOs purposively or accidentally enter into the human body, what will happen. Basically, a wider range of forager genetic materials can instill into the human alimentary tract everyday in a form different from materials that have never been reported to be problematic. This might be applied for the GMO however the nano GMO might have some differences. Due to the nanolevel, size effect should be carefully focused upon. In addition, the increase in reactivity at the nanolevel should also be a concern. Some interesting reports on carcinogenicity of nanomaterials are presented in Table 3.

Table 3. Reports on carcinogenicity of nanomaterials

Authors	Details
Ristovski et al. [56]	In this work, particle emissions from twelve buses, operating alternately on low sulfur (LS; 500 ppm) and ultralow sulfur (ULS; 50 ppm) diesel fuel, were monitored [56]. Ristovski et al. found that sulfur in diesel fuel was playing an important role in the formation of nanoparticles and the greatest influence of the fuel sulfur content was observed at the highest engine load [56].
Memisoglu-Bilensoy et al. [57]	Cytotoxic evaluation of injectable cyclodextrin nanoparticles was studied [57]. In this work, nanoparticles were treated using beta-CDC6, which is an amphiphilic beta-cyclodextrin derivative modified on the secondary face with 6C aliphatic esters [57]. Memisoglu-Bilensoy et al. said that there was a relationship between surfactant existence and nanoparticle characteristics, however, these effects are not significant [57]. Memisoglu-Bilensoy et al. also found that surfactant existence has no effect on model medicine nimodipine encapsulation but accelerates the in-vitro release of the medicine [57].
Brunner et al. [58]	Brunner et al assessd a human mesothelioma and a rodent fibroblast cell line for in vitro cytotoxicity tests using seven experimentally important nanoparticles [58]. In this work, the nanoparticles' response in terms of metabolic activity and cell proliferation of cultures exposed to 0-30 ppm nanoparticles was compared to the effects of nontoxic amorphous silica and toxic crocidolite asbestos [58]. Brunner et al. a nanoparticle-specific cytotoxic mechanism for uncoated iron oxide and partial detoxification or recovery after treatment with zirconia, ceria, or titania [58].

Authors	Details
Duffus [59]	Carcinogenicity classification of vanadium pentoxide and inorganic vanadium compounds was studied in this work [59]. Duffus concluded that for all particulates, such as those of crystalline vanadium pentoxide, the special toxicity of particulates, and especially nanoparticles, required separate consideration and classification [59].
Roser et al. [60]	The surface charges on biodegradable albumin nanoparticles were introduced by covalent coupling different primary amines to examine their influence on phagocytosis by macrophages under in vitro conditions in this work [60]. Roser et al. concluded that no differences in blood circulation times and organ accumulation between different nanoparticle preparations with positive, neutral and negative surface charges could be observed in rats, suggesting that the in vivo fate of albumin nanoparticles was significantly influenced by factors not reflected in the in vitro cell culture models [60].

B. Teratogenicity

Teratogenicity is another important focus. There is limited knowledge about the area. It is necessary to perform further studies on teratogenicity of nanomaterial for ful control of the nanoproducts.

As a conclusion, the environmental impact of nanomaterials is a focus at present. Safety of nanomaterials requires further verification. Many seminal reports have identified gaps in our knowledge, and a large multidisciplinary effort will be required to undertake the necessary research to assist the framing of regulatory models to deal with any new risks [61]. Swidwinska-Gajewska said that exposure to engineered nanoparticles present in the sector of research studies and advancement of nanotechnology, in chemical, pharmaceutical, and cosmetics industries as well as in processes involving nanoparticles in the form of semi-products, has become a real threat [62]. Swidwinska-Gajewska also noted that an increased surface area per mass unit, characteristic of nanostructures, substantially influenced their toxicity [62]. Powers et al. said that recommendations regarding how best to approach nanomaterial characterization should include using proper sampling and measurement methods, forming multidisciplinary teams, and making measurements as close to the biological action point as possible [63].

Ethical Aspect of Nanotechnology

A useful method for control of scientific technology is ethical control. It is accepted that good ethical control should accompany scientific activity. Increasingly, risks from new and emerging technologies are being regulated at the international level, although governments and private experts are only beginning to consider the appropriate international responses to nanotechnology [64]. There are many aspects of ethical concern on new technology. The first basic principle is that it is necessary to establish that the new technology is an actually good technology. This also implies that it is a safe technology and does not have side effects. This fact was once discussed in the environmental aspect of nanotechnology in the previous part of this chapter. It should be noted that since the nanoparticle is still under study for its safety and there is a possibility of side effects, there is a necessity to set a strict control program for the nanomaterial synthesized in a laboratory. The synthesis laboratory must be accredited and set up for the safety of students, including units for physical bio-chemical aspects. Concept of ethics in laboratory medicine can be applied for nanotechnology [65 – 66]. The concept can also be applied for nano-, bio-, info-, and cogno- convergence technologies [67]. Sententia et al. proposed that scientists required analytical models that protect values of personhood at the heart of a functional democracy – values that allow, as much as probable, for individual decision-making, despite transformations in understanding and ability to manipulate cognitive processes [68]. Sententia et al. stated that addressing cognitive enhancement from the legal and ethical notion of cognitive liberty could bring a powerful tool for assessing and encouraging nano-bio-info-cogno developments [68]. It can be said that nanoethics is a new field of ethical inquiry [69]. The important topics in nanoethics will be further discussed.

A. Concern on Nanotechnology Research

Similar to any biomedical research, the main question for nanotechnology research is "How should the dignity of people participating in nanomedicine research experiments be respected?" The first main area to be considered is the protection of individuals involved in the nanotechnology research. Protection of individuals should be applied to both health care and medical research. The second concern is informed consent. This is the basic requirement in biomedical research. However, it may be difficult to bring adequate information concerning a proposed nanotechnology research because there is no warranty on the safety of nanoparticles as previously mentioned. Initiatives should be taken to enhance information exchange between research ethics committees in different settings. Although there are many methods for applied nanobiotechnology, health care concerns must be given first priority [70 – 77].

Table 4. Reports on ethicists' ideas on nanoethics for nanomedicine

Authors	Details
Lupton [71]	Lupton said that nanotech holds the potential to revolutionise medicine, electronics and chemistry, and nanomedicine would help the repair and improvement of the human body from the inside out, with a precision and delicacy far greater than the finest surgical tools allow [71]. Lupton quoted that problem areas stemming from the technology included the main questions "Who will benefit - just the rich or the poor too?" [71].
Resnik and Tinkle [74]	Resnik and Tinkle noted that nanomedicine revealed tremendous promise for improving medical diagnosis, treatment, and prevention, but it also increased a multiple appearance of ethical concerns [74]. Because of the paucity of data on the physicochemical properties of nanoscale materials in biological systems, Resnik and Tinkle proposed that clinical experiments of nanomedicine products presented some challenges related to risk minimization, management and communication involving human subjects [74].

Authors	Details
Best and Khushf [75]	Consider two ways that nanomedicine might be disruptive. Best and Khushf mentioned the two main concerns in naomedicine: a) low-end disruptions that were intrinsically unpredictable but limited in scope, and b) high-end disruptions that involved broader societal issues but could be anticipated, allowing opportunity for ethical reflection [75].
Lenk et al. [76]	Lenk et al. noted that ethics for nanomedicine included the issues of human identity and enhancement, probable implications for civil liberties, and concerns about equity and fair access [76]. Lenk et al. proposed that although it seemed that the main ethical issues were not unique to nanotechnologies, the conclusion would argue against shrugging them off as non-specific items that had been considered before in the context of other biomedical technologies, such as gene therapy or xenotransplantation [76].
Khushf [77]	Khushf considers two types of nanomedicine and the kinds of ethical issues they raised were a) nanomedicine of an incremental kind that requires proper ethical assessment, and b) nanomedicine of a more programmatic kind [77]. Khushf noted that general concepts of medical ethics can be applied for the first kind [77]. For the second kind, Khushf noted that more programmatic developments included challenges integral to formation of interdisciplinary teams; issues related to intellectual property, authorship and publication; development of informed consent and confidentiality protections associated with new data sets; future challenges to the clinician-patient relation and personalized medicine should be concerns [77].

B. Concern on Environment

"How can we protect the fundamental rights of citizens that may be exposed to free nanoparticles in the environment?" is a query that needs an answer. As a social responsibility, the synthesized laboratory must have a well-controlled system and obtain international standards for analysis and safety. Responsible use of nanomedicine which protects both human health and the environment needs to be established.

LEGAL ASPECT OF NANOTECHNOLOGY [72]

The legal aspect of nanotechnology is a new concern. Because it is a new thing, there might be an absence of specific nanotechnology law. However, the legal basic principles for science can be applied. Basically, in the West, the legal system includes legislation on products, clinical experiments, data and patents, medicinal products, aspects of intellectal property, medical devices, cosmetics, chemicals, as well as data protection. These issues should be addressed in the producing world. Data protection and patent directives should be focused. In addition, environmental control law should also be reviewed for nanotechnology. Miller noted that a responsible government agency should consider implementing many reforms to make sure that it was adequately treated to regulate nanobiotechnology [78].

REFERENCES

[1] Kane RS, Stroock AD. Nanobiotechnology: protein-nanomaterial interactions. *Biotechnol. Prog.* 2007 Mar-Apr;23(2):316-9.

[2] Medvedeva NV, Ipatova OM, Ivanov IuD, Drozhzhin AI, Archakov AI. Nanobiotechnology and nanomedicine. *Biomed. Khim.* 2006 Nov-Dec;52(6):529-46.

[3] Blattler T, Huwiler C, Ochsner M, Stadler B, Solak H, Voros J, Grandin HM Nanopatterns with biological functions. *J. Nanosci. Nanotechnol.* 2006 Aug;6(8):2237-64.

[4] Groneberg DA, Giersig M, Welte T, Pison U. Nanoparticle-based diagnosis and therapy. *Curr .Medicine Targets.* 2006 Jun;7(6):643-8.

[5] Jain KK. The role of nanobiotechnology in medicine discovery. *Medicine Discov. Today.* 2005 Nov 1;10(21):1435-42.

[6] Penn SG, He L, Natan MJ. Nanoparticles for bioanalysis. *Curr. Opin. Chem. Biol.* 2003 Oct;7(5):609-15.

[7] Laval JM, Mazeran PE, Thomas D. Nanobiotechnology and its role in the development of new analytical devices. *Analyst.* 2000 Jan;125(1):29-33.

[8] Swidwinska-Gajewska AM. Nanoparticles (part 1)--the product of novel technology and new hazards in the work environment. *Med. Pr.* 2007;58(3):243-51.

[9] Zuo L, Wei W, Morris M, Wei J, Gorbounov M, Wei C. New technology and clinical applied usages of nanomedicine. *Med. Clin. North Am.* 2007 Sep;91(5):845-62.

[10] Murthy SK. Nanoparticles in novel medicine: state of the art and future challenges. *Int. J. Nanomedicine.* 2007;2(2):129-41.

[11] Brewer M, Zhang T, Dong W, Rutherford M, Tian ZR. Future approaches of nanomedicine in clinical science. *Med. Clin. North Am.* 2007 Sep;91(5):963-1016.

[12] Morrow KJ Jr, Bawa R, Wei C. Recent advances in basic and clinical nanomedicine. *Med. Clin. North Am.* 2007 Sep;91(5):805-43.

[13] Emerich DF. Nanomedicine--prospective therapeutic and diagnostic applied usages. *Expert Opin. Biol .Ther.* 2005 Jan;5(1):1-5.

[14] Medvedeva NV, Ipatova OM, Ivanov IuD, Drozhzhin AI, Archakov AI. Wei C, Wei W, Morris M, Kondo E, Gorbounov M, Tomalia DA. Nanomedicine and medicine delivery. Med Clin North Am. 2007 Sep;91(5):863-70. Nanobiotechnology and nanomedicine. *Biomed. Khim.* 2006 Nov-Dec;52(6):529-46.

[15] Mustelin T. Challenges and optimism for nanoengineering. *Nanomed.* 2006 Dec;1(4):383-5.

[16] Wikswo JP. The need for cellular and molecular sensors and actuators. Conf Proc IEEE *Eng. Med. Biol. Soc.* 2004;7:5393.

[17] Finlay WH, Moussa W. Medical and pharmaceutical nanoengineering conference/International conference on MEMS, nano and smart systems.Expert Opin *Medicine Deliv.* 2004 Nov;1(1):177-80.

[18] Liu XY, Sawant PD. Micro/Nanoengineering of the self-organized three-dimensional fibrous structure of functional materials. Angew *Chem. Int. Ed. Engl.* 2002 Oct 4;41(19):3641-5, 3518.

[19] Maysinger D. Nanoparticles and cells: good companions and doomed partnerships. *Org. Biomol. Chem.* 2007 Aug 7;5(15):2335-42

[20] Kailasapathy K. Microencapsulation of probiotic bacteria: technology and potential applied usages. *Curr .Issues Intest. Microbiol.* 2002 Sep;3(2):39-48

[21] Broberg A, Thomsen KK, Duus JO. Applied usage of nano-probe NMR for structure determination of low nanomole summations of arabinoxylan oligosaccharides fractionated by analytical HPAEC-PAD. *Carbohydr. Res.* 2000 Sep 22;328(3):375-82.

[22] Pfenninger A, Karas M, Finke B, Stahl B. Structural analysis of underivatized neutral human milk oligosaccharides in the negative ion mode by nano-electrospray MS(n) (part 2: applied usage to isomeric mixtures). *J. Am. Soc. Mass. Spectrom.* 2002 Nov;13(11):1341-8.

[23] von Bultzingslowen C, McEvoy AK, McDonagh C, MacCraith BD, Klimant I, Krause C, Wolfbeis OS. Sol-gel based optical carbon dioxide sensor employing couple luminophore referencing for applied usage in food packaging technology. *Analyst.* 2002 Nov;127(11):1478-83.

[24] Cataldi TR, Nardiello D, Ciriello R, Guerrieri A. Pulsed electrochemical detection of orotic acid by an activated potential waveform at a gold working electrode following anion-exchange chromatography. *J. Chromatogr.* A. 2006 Feb 24;1107(1-2):130-8.

[25] Chen J, Liu M, Zhang L, Zhang J, Jin L. Applied usage of nano TiO(2) towards polluted hydro treatment combined with electro-photochemical method. *Hydro Res.* 2003 Sep;37(16):3815-20.

[26] Kwon IK, Kidoaki S, Matsuda T. Electrospun nano- to microfiber fabrics made of biodegradable copolyesters: structural characteristics, mechanical properties and cell adhesion potential. *Biomaterials.* 2005 Jun;26(18):3929-39.

[27] Dan N. Synthesis of hierarchical materials. Potentials Biotechnol. 2000 Sep;18(9):370-4.

[28] Erickson RP. From "magic bullet" to "specially engineered shotgun loads": the new genetics and the need for individualized pharmacotherapy. *Bioessays.* 1998 Aug;20(8):683-5.

[29] Osieka R. Oncology. *Arzneimittelforschung.* 1997 Oct;47(10):1161-5.

[30] Galanski M, Keppler BK. Searching for the magic bullet: anticancer platinum medicines which can be accumulated or activated in the tumor tissue. *Anticancer Agents Med. Chem.* 2007 Jan;7(1):55-73.

[31] Welch DR. Biologic considerations for medicine targeting in cancer patients. *Cancer Treat. Rev.* 1987 Dec;14(3-4):351-8.

[32] Kreuter MW. Human property and cancer. Forobtain the magic bullet! *Cancer.* 1993 Aug 1;72(3 Suppl):996-1001.

[33] Couvreur P, Vauthier C. Nanotechnology: intelligent design to treat complex disease. *Pharm. Res.* 2006 Jul;23(7):1417-50.

[34] Ullman D. Let's have a serious discussion of nanopharmacology and homeopathy. *FASEB J.* 2006 Dec;20(14):2661; author reply 2662. No

[35] Yokoyama M.Medicine targeting with nano-sized carrier systems. *J Artif Organs.* 2005;8(2):77-84.

[36] Sahoo SK, Labhasetwar V. Nanotech approaches to medicine delivery and imaging. *Medicine Discov Today.* 2003 Dec 15;8(24):1112-20.

[37] Roco MC. Science and technology integration for increased human potential and societal outcomes. *Ann. N. Y. Acad. Sci.* 2004 May;1013:1-16.

[38] Spohrer JC, Engelbart DC. Converging technologies for enhancing human characteristic: science and business perspectives. *Ann. N. Y. Acad. Sci.* 2004 May;1013:50-82.
[39] Brumfiel G. Consumer products leap aboard the nano bandwagon. *Nature.* 2006 Mar 16;440(7082):262.
[40] Joachim C. To be nano or not to be nano? *Nat. Mater.* 2005 Feb;4(2):107-9.
[41] Service RF. Science policy. Priorities needed for nano-risk research and development. *Science.* 2006 Oct 6;314(5796):45.
[42] Robichaud CO, Tanzil D, Weilenmann U, Wiesner MR. Relative risk analysis of many manufactured nanomaterials: an insurance industry context. *Environ. Sci. Technol.* 2005 Nov 15;39(22):8985-94.
[43] Stern ST, McNeil SE. Nanotechnology Safety Concerns Revisited. *Toxicol. Sci.* 2007 Jul 4;
[44] Bowman DM, Fitzharris M. Too small for concern? Public health and nanotechnology. *Aust. N Z J. Public Health.* 2007 Aug;31(4):382-4.
[45] Allen TM, Cheng WW, Hare JI, Laginha KM. Pharmacokinetics and pharmacodynamics of lipidic nano-particles in cancer. *Anticancer Agents Med. Chem.* 2006 Nov;6(6):513-23.
[46] Rigdon RH. Plastics and carcinogenesis. *South Med. J.* 1974 Dec;67(12):1459-65.
[47] Kichler A.Gene transfer with modified polyethylenimines. *J. Gene Med.* 2004 Feb;6 Suppl 1:S3-10.
[48] El-Aneed A. An overview of present delivery systems in cancer gene therapy. *J. Control. Release.* 2004 Jan 8;94(1):1-14.
[49] Lungwitz U, Breunig M, Blunk T, Gopferich A. Polyethylenimine-based non-viral gene delivery systems. *Eur. J. Pharm. Biopharm.* 2005 Jul;60(2):247-66.
[50] Lemkine GF, Demeneix BA. Polyethylenimines for in vivo gene delivery. *Curr. Opin. Mol. Ther.* 2001 Apr;3(2):178-82.
[51] Godbey WT, Wu KK, Mikos AG.Poly(ethylenimine) and its role in gene delivery. *J. Control Release.* 1999 Aug 5;60(2-3):149-60.
[52] Abdallah B, Sachs L, Demeneix BA. Non-viral gene transfer: applied usages in developmental biology and gene therapy. *Biol. Cell.* 1995;85(1):1-7.
[53] Ciliberti R, Molinelli A. Towards an GMO discipline: ethical remarks.*Vet. Res. Commun.* 2005 Aug;29 Suppl 2:27-30.
[54] Schauzu M. Genetically modified plants and food safety. State of the art and discussion in the European Union. *Bundesgesundheitsblatt Gesundheitsforschung Gesundheitsschutz.* 2004 Sep;47(9):826-33.
[55] Yang L, Yang X. Progress on biosafety assessment of marker genes in genetically modified foods. *Wei Sheng Yan Jiu.* 2003 May;32(3):239-45.
[56] Ristovski ZD, Jayaratne ER, Lim M, Ayoko GA, Morawska L. Influence of diesel fuel sulfur on nanoparticle emissions from city buses. *Environ. Sci. Technol.* 2006 Feb 15;40(4):1314-20.
[57] Memisoglu-Bilensoy E, Dogan AL, Hincal AA. Cytotoxic evaluation of injectable cyclodextrin nanoparticles. *J. Pharm. Pharmacol.* 2006 May;58(5):585-9.
[58] Brunner TJ, Wick P, Manser P, Spohn P, Grass RN, Limbach LK, Bruinink A, Stark WJ. In vitro cytotoxicity of oxide nanoparticles: comparison to asbestos, silica, and the effect of particle solubility. *Environ. Sci. Technol.* 2006 Jul 15;40(14):4374-81.

[59] Duffus JH. Carcinogenicity classification of vanadium pentoxide and inorganic vanadium compounds, the NTP study of carcinogenicity of inhaled vanadium pentoxide, and vanadium chemistry. *Regul. Toxicol. Pharmacol.* 2007 Feb;47(1):110-4.

[60] Roser M, Fischer D, Kissel T. Surface-modified biodegradable albumin nano- and microshperical objects. II: effect of surface charges on in vitro phagocytosis and biodistribution in rats. *Eur. J. Pharm. Biopharm.* 1998 Nov;46(3):255-63.

[61] Priestly BG, Harford AJ, Sim MR. Nanotechnology: a useful new technology--but how safe? *Med. J. Aust.* 2007 Feb 19;186(4):187-8.

[62] Swidwinska-Gajewska AM. Nanoparticles (part 1)--the product of novel technology and new hazards in the work environment] *Med. Pr.* 2007;58(3):243-51.

[63] Powers KW, Brown SC, Krishna VB, Wasdo SC, Moudgil BM, Roberts SM. Research strategies for safety evaluation of nanomaterials. Part VI. Characterization of nanoscale particles for toxicological evaluation. *Toxicol. Sci.* 2006 Apr;90(2):296-303.

[64] Marchant GE, Sylvester DJ. Transnational models for regulation of nanotechnology. *J Law Med Ethics*. 2006 Winter;34(4):714-25.

[65] Wiwanitkit V. Ethics of clinical pathologist. *J. Med. Assoc. Thai.* 2006 Dec;89(12):2161-2.

[66] Resnik DB, Tinkle SS. Ethics in nanomedicine. *Nanomed.* 2007 Jun;2(3):345-50.

[67] Khushf G. Systems theory and the ethics of human enhancement: a framework for NBIC convergence. *Ann. N. Y. Acad. Sci.* 2004 May;1013:124-49.

[68] Sententia W. Neuroethical considerations: cognitive liberty and converging technologies for improving human cognition. *Ann. N. Y. Acad. Sci.* 2004 May;1013:221-8.

[69] Grunwald A. Nanotechnology--a new field of ethical inquiry? *Sci Eng Ethics*. 2005 Apr;11(2):187-201.

[70] Bawa R, Johnson S. The ethical dimensions of nanomedicine. *Med. Clin. North Am.* 2007 Sep;91(5):881-7.

[71] Lupton M. Nanotechnology--salvation or damnation for humans? *Med Law.* 2007 Jun;26(2):349-62.

[72] Hermeren G. Nanomedicine challenges European ethicians, lawyers and toxicologists. *Lakartidningen*. 2007 Apr 25-May 1;104(17):1326-30.

[73] Thompson RE. Nanotechnology: science fiction? Or next challenge for the ethics committee? *Physician Exec.* 2007 May-Jun;33(3):64-6.

[74] Martin-Lomas M. Bio-nanoscience and bio-nanotechnology: social and ethical aspects. *Law Hum. Genome Rev.* 2006 Jul-Dec;(25):91-107.

[75] Best R, Khushf G. The social conditions for nanomedicine: disruption, systems, and lock-in. *J. Law Med. Ethics.* 2006 Winter;34(4):733-40.

[76] Lenk C, Biller-Andorno N. Nanomedicine-emerging or re-emerging ethical issues? A discussion of four ethical themes. *Med. Health Care Philos.* 2007 Jun;10(2):173-84.

[77] Khushf G. Upstream ethics in nanomedicine: a call for research. *Nanomed.* 2007 Aug;2(4):511-21.

[78] Miller J. Beyond biotechnology: FDA regulation of nanomedicine.*Columbia Sci. Technol. Law Rev.* 2003;4:E5.

Chapter 15

QUANTUM MEDICINE

INTRODUCTION TO QUANTUM

Quantum is a physical term that means the smallest discrete quantity of some physical property that a system can possess. Quantum theory, the branch of physics which is based on quantization, began in 1900 when Planck published his theory explaining the emission spectrum of black bodies [1 - 3]. The biphasic of radiation and matter is the heart of the quantum theory. In physics, quantum mechanics is the study of the relationship between energy or radiation and matter, in particular that between valence shell electrons and photons. Bohr's theory is the root for the development of the present quantum physics [4 - 5]. Bohr's atomic theory is generally viewed as remarkable, both for its accuracy in predicting the observed optical transitions of one-electron atoms and for its failure to completely correspond with present electronic structure theory [6]. However, what is not generally appreciated is that Bohr's starting point of a semiclassical conception differed significantly from the Bohr-Sommerfeld theory and offers an alternative semiclassical approximation scheme with remarkable attributes [6]. It is generally believed that the old quantum theory, as presented by Bohr in 1913, fails when applied to a few electron systems, such as the H(2) molecule [7]. Bohr's starting point method did not impose action quantization constraints but rather obtained these as predictions by simply matching photon and classical orbital frequencies [6]. The discovery of the efficacy-quantum by Planck and further work of Bohr was the basic contributing root to Einstein's hypothesis of lightquanta [8]. With the mentioned basic theories, Einstein could explain the external light-electrical effect [8]. The work of Einstein is the root of the new phase of science. For example, it was a long way from the concept of stimulated emission as the fundamental idea of laser technology by Albert Einstein in 1917 to the practical use of the laser today [9]. It is also the basis for the new science, quantum mechanics. Quantum mechanics is a fundamental branch of physics with wide applied uses in both experimental and theoretical physics [10- 13]. The quantum theory is the theoretical basis of modern physics that can be useful in explaining the nature and property of matter and energy on the atomic and subatomic level. Conrad said that biological cells had greater information processing efficiency than the programmable computers used to model them [12]. Conrad also suggested that systems in which quantum features play a prominent role are more powerful than classical physical-dynamical analogs [12].

WHAT IS QUANTUM MEDICINE?

Quantum medicine is the new generation of medicine. Naturally, all things, chemicals, plants, animals as well as human beings are composed of various tissues, organs, cells, molecules, and atoms. The combinations of atoms result in molecules. The combinations of molecules result in cells. The combinations of cells results in tissues. The combinations of tissues result in organs. The combinations of organs result in human beings. Basically, an atom consist of electrons and a nucleus which includes the intern quantum and neutron. Electrons have a negative charge, protons have a positive charge and neutrons have a neutral charge. Normally human beings cannot see the atom. In addition, the compositions of atoms are beyond human reach. But according to the new elementary particle theory, the elementary particle is considered to have immense size and types, and it has the property of an energy field which rotates and vibrates itself. With the advent of nanoscience, those very tiny compositions can be studied.

Quantum measurement method is the measurement way, which tests the energy composed of magnetic energy produced by quantum and electron motion. Improvements in quantum chemical methods have led to increased applied usages to biological problems, including the development of potential energy functions for molecular mechanics and modeling of the reactive chemistry in enzyme active sites, with particularly interesting progress being made for metal-containing systems [13]. The three areas in which the quantum measurement method is used include (a) use of quantum mechanical electronic structure methods such as molecular orbital theory and density functional theory, usually in conjunction with molecular mechanics; (b) treating vibrational movements quantum mechanically, either in an instantaneous harmonic approximation, or by path integrals, or by a three-dimensional wave function coupled to classical nuclear motion; (c) incorporation of multidimensional tunneling approximations into reaction rate calculations [14]. Orbital theory is the basic knowledge for quantum chemistry. Molecular orbital analysis is the basic way for study of quantum mechanical electronic structure. To help molecular orbital analysis, there are many new computation tools for performing this analytical purpose. For example, AOMix is a new software for the molecular orbital analysis [15]. It calculates the molecular orbital compositions in terms of the constituent chemical fragments in the molecule or atom [15]. AOMix automatically processes output files of multiple quantum-chemical packages [15]. AOMix also allows one to analyze chemical structure using overlap populations (total and per molecular orbital), valence indices, 2-center (Mayer, Lowdin, Wiberg, and bond-order symmetry components) and 3- and 4-center bond orders, charge decomposition orbital, CDA (total, per molecular orbital, per symmetry type), and condensed Fukui functions [15]. Conventional quantum chemistry methods are either too expensive to apply to large systems or too approximate for the results to be reliable, and they fail to satisfy this requirement [16]. A multiple appearance of different approaches is being developed with the aim of achieving this goal: local correlation methods; divide-and-conquer methods; linear-scaling density functional methods based on the fast multipole and other approximations; effective potential methods; and hybrid methods [16].

The influence and applied usage of quantum measurement in the medical field is called "quantum medicine". Quantum medicine is one of the most useful areas of novel contemporary medicine [17 – 18]. It is based on the use of purposeful effects of low-dosage electromagnetic radiation or quantum in treatment, diagnosis, prophylaxis, and rehabilitation of patients.

QUANTUM DIAGNOSIS

Applied usage of quantum medicine diagnosis is a main focus. There are many attempts to develop a new diagnostic tool at quantum level [19 - 20]. Quantum diagnosis relies on energy and information parameters that specifically depict both the general state of the organism and its organs, systems and interrelations. Measurements of relevant electric parameters at certain points of the body surface allow one to diagnose for a disease or to identify it at the early onset stage with high accuracy. Measurements of those parameters are also helpful to define predisposition of the patient to specific disease. In nanomedicine, a well-known diagnostic tool, namely quantum dot, is now available. Quantum dot is a synthesized nanopolymer with semiconductor property. Quantum dot is a new class of inorganic fluorophore which is gaining widespread interest as a result of their exceptional photophysical properties. Semiconductor quantum dots are inorganic nanoparticles with unique photophysical properties [21]. In particular, their huge one- and two-photon absorption cross sections, tunable emission bands and excellent photobleaching resistances are stimulating the development of luminescent probes for biomedical imaging and sensing applied usages [21]. With the nanoproperty of quantum dot, electron and energy transfer processes can be designed to interprete the luminescence of semiconductor quantum dots in response to molecular recognition events [21]. On the basis of these operating principles, the existence of target analytes can be transduced into detectable luminescence signals. Quantum dot is generalized applied used for staining of living cells. The quantum dot is more stable than classical fluorescence staining, which was accepted as one of the most useful candidates for chemical sensing [22]. At present, quantum dots have been increasingly employed in measuring the dynamic property of biomacromolecules using fluorescence correlation spectroscopy [23]. This brings a challenge, because quantum dots display their own dynamic property in the form of intermittent photoluminescence, also known as blinking [23].

Bioconjugated nanoparticles and quantum dots are among the most exciting nanomaterials with useful applied usage potentials in the nanomedicine field [24]. Although most of these applied usages are based on the optical properties of nanoparticle materials such as surface plasmon resonance, surface-amplified Raman scattering and strong photoluminescence, other aspects of nanoparticles such as the catalytic effect and amplification effect associated with the nanoscale dimension have also been explored [24]. In addition to standard quantum dot, now functional antibodies, antibodies that are sterically available for functional binding to target proteins can be completely conjugated to semiconductor quantum dots [25]. This is a new potential method for improving quantum dot tagging of biological preparations [25]. However, Pathak et al. found that the number of available functional antibodies varied significantly for different conjugation methods and were lower than previously estimated [26]. Pathak et al. said that these results might suggest potential strategies for improving quantum dot tagging of biological preparations [26]. Interesting reports on the applied usage of quantum dot on this area are listed in Table 1.

Table 1. Examples of reports on the antibody conjugated quantum dot for diagnosis in medicine

Authors	Details
Chan and Nie [27]	Chan and Wu reported the in development of highly luminescent semiconductor quantum dots (zinc sulfide-capped cadmium selenide) covalently coupled to biomolecules for use in ultrasensitive biological detection [27]. According to this work, in comparison with organic dyes such as rhodamine, this class of luminescent labels was 20 times as bright, 100 times as stable against photobleaching, and one-third as wide in spectral linewidth [27].
Goldman et al. [28]	Goldman et al. described the preparation and characterization of bioinorganic conjugates in which luminescent semiconductor CdSe-ZnS core-shell nanocrystal quantum dots coupled to antibodies through the use of an avidin bridge adsorbed to the nanocrystal surface via electrostatic self-assembly [28]. Goldman et al. noted that quantum dot conjugation to biotinylated antibodies then was readily getd and fluoroimmunoassays utilizing these antibody conjugated quantum dots were ful in the detection of protein toxins [28].
Tokumasu and Dvorak [29]	Tokumasu and Dvorak reported the first applied usage of quantum dots to immunocytochemical studies of human erythrocytes capable of being used in high-magnification, three-dimensional erythrocyte reconstruction ways [29]. According to this work, the procedure consist ofs of stabilizing human erythrocytes with a homofunctional imidoester cross-linker to decrease fixative-induced autofluorescence followed by reacting with a quantum dot - monoclonal antibody complex to label band 3 protein [29]. Tokumasu and Dvorak demonstrated the applicability of this way to studies of erythrocyte membrane modifications occurring during the invasion of a malaria parasite [29].
de Farias et al. [30]	de Farias et al. reported a new methodology for red blood cell antigen expression determination by a simple tagging procedure employing luminescent semiconductor quantum dots [30]. According to this work, fluorescence intensity maps revealed different antigen expressions for the distinct erythrocyte types [30]. de Farias suggested that this simple tagging procedure might be employed as an efficient tool to investigate quantitatively the distribution and expression of antigens in red blood cell membranes [30].
de Farias et al. [31]	de Farias et al. reported a new and efficient methodology of tagging red blood cells, in order to investigate the expression of anti-A antigen, employing luminescent semiconductor nanocrystals [31]. de Farias et al. suggested that this labelling procedure might be applied as a quantitative tool to investigate the distribution and expression of alloantigen in red blood cells [31].
Akhtar et al. [32]	Akhtar et al. described the use of quantum dot conjugates to detect primary antibody binding in fixed tissue sections [32]. Akhtar et al. also describe the use of quantum dots in simultaneous and sequential multitagging procedures and in combination with enzyme-based signal amplification ways [32]. Akhtar et al. said that quantum dot conjugates expanded the range of the immunohistochemist and increase experimental flexibility in many applied usages [32].

Authors	Details
Ma et al. [33]	Polystyrene fluorescent microshperical objects treated by deposition of CdTe quantum dots were used in an immunoassay in this study [33]. Ma et al. reported that this approach allowed detection of the antigen in the range 1-500 mg/L, based on the change in the fluorescence intensity of the reporter [33].
Dwarakanatha et al. [34]	In this study, the effect of a 20-min exposure to antibody-quantum dot conjugates on colony counts of *Escherichia coli* was assessed by the spread-plate method and compared with exposure to unconjugated quantum dots having only amine or carboxyl groups on their surfaces [34]. Under these conditions, antibody-quantum dot conjugates generally exhibited more than 90% reduction in colony-forming units as compared to untreated *E. coli* and E. coli treated with unconjugated quantum dots after incubation for as long as 41 hours [34].
Shen et al. [35]	In this work, a new monoclonal antibody against sulfamethazine was completely produced and linked to quantum dot nanocrystals by covalent coupling [35]. Shen et al. said that determination of sulfamethazine in milk by fluoroimmunoassay was feasible [35].
Wang et al. [36]	In this work, different diameter CdSe/ZnS semiconductor nanocrystals, quantum dots were synthesized by changing the nucleation time, using organometallic reagents [36]. Wang et al. reported that the encoded beads could detect the antibody signal effectively and the effective detection range of the antibody is about 2-15 muM [36].
Liu et al. [37]	Liu et al. described a disposable electrochemical immunosensor diagnosis device that integrates the immunochromatographic strip way with an electrochemical immunoassay and exploits quantum dot as labels for amplifying signal output [37]. Liu et al. said that the new disposable electrochemical diagnosis device could bring a rapid, clinically accurate, and quantitative tool for protein biomarker detection [37].

In addition to antibody, conjugation can be made with the use of other substances. Another generalized used substance is peptide. In 2005, Tomlinson et al. treated peptide-quantum dot conjugates by attaching angiotensin II (Ang II) to cadmium selenide/zinc sulfide core-shell nanocrystals using an 1-[3-(Dimethyamino)propyl]-3-ethylcarbo diimide hydrochloride (EDC) coupling [38]. According to this work, the Ang II-quantum dot conjugates did not bind to parental cells and revealed similar staining patterns when compared with a commercially available conjugate [38]. Young and Rozengurt demonstrated that quantum dot-ANG II was brighter and more photostable than agonist labeled with the organic dye Cy3 [39]. Young and Rozengurt demonstrated that quantum dot technology could be adapted to monitor ligand binding to G protein-coupled receptors [39]. In 2007, Tsay et al. reported the study on singlet oxygen production by peptide-coated quantum dot-photosensitizer conjugates [40]. Tsay et al found that the conjugate could be simultaneously used for fluorescence imaging and singlet oxygen generation [40]. In addition, singlet oxygen quantum yields as high as 0.31 were getd using 532-nm excitation wavelengths [40].

Quantum dot can also be applied for imaging [41]. The new generations of quantum dots have far-reaching potential for the study of intracellular processes at the single-molecule level, high-resolution cellular imaging, long-term in vivo observation of cell trafficking, tumor targeting, and diagnostics [41]. Quantum dots are emerging as a superior alternative and are poised to change the world of bio-imaging and further its applied usages in basic and applied biology [42 - 46]. Jaiswal and Simon said that the improved synthesis of hydro-stable quantum dots, the development of approaches to label cells efficiently with quantum dots, and improvements in conjugating quantum dots to specific biomolecules had triggered the recent growing in their use in biological imaging [47]. Jaiswal and Simon also noted that although there had been many es in using quantum dots for biological applied usages, limitations remained that required solution before these powerful tools could be used routinely by biologists [47].

QUANTUM TREATMENT

Quantum treatment is a treatment based on quantum [48 – 49]. Quantum treatment recruits natural, environment-like factors of electromagnetic exposure that have beneficial effects on cells, organs and systems of the organism. It is accepted that the low level of energy used in quantum medicine is absolutely harmless. The treatment reverses existing anomalous process, disease, to stability and health. This is because quantum treatment activates latent reserves of the organism adaptation both at the cellular level and whole biologic system, strongly enhancing immunity and mobilizing the organism for active resistance to the functional disorder. Quantum medicine is increasingly applied for treatment in all fields of medicine. It is generalized used in cases of cancer, radiation sickness, cardiovascular diseases, locomotors diseases, infantile cerebral palsy, urolithiasis, bronchial asthma, alopecia, enuresis and infertility [50 – 56]. Optical quantum generators or the laser is the best example of quantum treatment. The optical quantum generators are useful in orthopedics/ traumatology, genecology, urology, sports medicine, dermatology, dentistry, surgery, neurology and rheumatology [50 – 56]. Indeed, the quantum therapy has a long history. The classical well-known therapy is the phototherapy in infantile hyperbilirubinemia [57]. In phototherapy, the structural photoisomerization is generally assumed to represent the main route of bilirubin elimination [58]. As a consequence, the determination of the spectral band that optimizes the process of formation of lumirubin in the neonate may represent an important step in the improvement of the clinical protocol of phototherapy [58]. The laser therapy is another classical treatment as previously mentioned.

Besides the treatment by the classical quantum medical tool, the medicine treatment for general diseases is also based on the principle of quantum change. To make the better understand this topic, the author hereby quotes interesting reports describing the quantum mechanism in treatment of some important diseases.

Based on the advent of nanobiotechnology, quantum dot can also be applied for quantum treatment. At present, the photophysics of singlet oxygen generation in relation to quantum dot-based energy transfer is generalized discussed and the possibility of using quantum dots as photosensitizer in photodynamic therapy is generalized assessed, including their present limitations to applied usages in biological systems [62 - 63]. Basically, photodynamic therapy is an emerging treatment modality for a range of disease classes, both cancerous and noncancerous [64]. This has brought about an active pursuit of new photodynamic therapy agents that can be optimized for the unique set of photophysical characteristics that are required for a ful clinical agent [64 - 65]. Hsieh et al. reported the in design and synthesis of Ir-complex functionalized CdSe/ZnS quantum dots, in which the QD played a key role in imaging, while the Ir-complex acted as a sensitizer to produce singlet oxygen for a photodynamic therapy purpose [66].

Table 2. Reports describing quantum mechanism in treatment of some important diseases

Authors	Details
Wiwanitkit [59]	In this work, Wiwanitkit performed a quantum chemical analysis of the activation reaction of artemisinin by the two alternative pathways [59]. According to this study, the required energy for compound formation in C-C cleavage is more than that for C-O cleavage [59]. It can be noted that the C-C cleavage pathway is less preferable, implying that the 1.5 H-shift should be the more common phenomenon [59].
Wiwanitkit [60]	In this work, a quantum chemical analysis of the deferiprone-iron binding reaction was performed, focusing on the reaction between one molecule of deferiprone and one molecule of iron [60]. The two main alternative pathways for the deferiprone-iron binding reaction were shown to be C-C cleavage and C-O cleavage [60]. The required energy for complex formation in C-C cleavage was less than for C-O cleavage [60]. The total energy requirement for C-C cleavage was negative, implying that this reaction can occur without any external energy source [60]. Wiwanitkit reported that the resulting complex fit the reported tertiary structure model for the deferiprone-iron complex [60].
Wiwanitkit [61]	Wiwanitkit performed an analysis on binding energy activity of TIBO and HIV-RT based on simple consideration for conformational change [61]. According to this work, the change of binding energy between TIBO and HIV-RT due to the variation in beta rotatable dihedral angles was detected [61]. The theoretical simulation for a step size of 30 degrees variation from 0 to 90 degree changes was performed [61]. The derived binding energies range from 0 to 8 kcal/mol [61]. Changes in the beta dihedral angle conformation critically impacts the binding energy. TIBO derivatives that impact the beta angle can change the binding energy between medicine and HIV-RT and this can impact medicine activity [61].

Quantum Prophylaxis and Quantum Rehabilitation

In addition to therapy, quantum prophylaxis can also be performed based on the methods of quantum therapy. For instance, prophylaxis of disease relapses in periodic exacerbation requires quantum treatment at the early stage of an epidemic. There are also some interesting reports on applied usage of quantum dot in vaccinology. Upadhyay investigated how mild local hyperthermia could be used for transdermal-immunization [67]. Upadhyay said that hyperthermia amplified transdermal-immunization procedure was likely to have higher compliance as it did not cause any pain or visible damage to the skin [67].

Quantum rehabilitation also makes use of quantum therapy methods. The use of quantum medicine technologies can help fasten patient rehabilitation period after complex operations, radiation or chemical exposures. The most generalized used quantum rehabilitation is for cardiological problem [68 - 69]. The quantum gravitational therapy is the generalized mentioned method. This way is a complex therapy of myocarditis of rheumatic and non-rheumatic genesis using intravascular laser irradiation of blood, quercitrol, and enterosgel; has an antiinflammatory, antioxidant action, improves myocardial contractility, endowed with an antiaggregatory activity [68]. This way is warranted for its effectiveness in rehabilitation in myositis. Nevertheless, Hossu and Rupert said that biophoton emission was a quantum event characterized by a relatively stable but ultralow-rate emission of visible photons from living organisms, and it has been associated with high-energy processes such as cell metabolism, growth, phagocytosis, neural activity, and oxidative stress [70]. Hossu and Rupert concluded that biophoton intensity was a noninvasive monitoring parameter for the health of the human body and is significantly altered in different ways by chiropractic interventions [70].

REFERENCES

[1] Shampo MA, Kyle RA. Max Planck. *JAMA.* 1980 Nov 28;244(21):2429.

[2] Cardinale AE. An adventure of the human intellect. Max Planck and the centenary of quantum theory. *Radiol. Med.* (Torino). 2000 Mar;99(3):127-34.

[3] Staley R. On the co-creation of classical and novel physics. *Isis.* 2005 Dec;96(4):530-58.

[4] Shampo MA, Kyle RA. Niels Bohr. *JAMA.* 1975 Sep 1;233(9):992.

[5] del Regato JA. Niels Bohr. Int J Radiat *Oncol. Biol. Phys.* 1981 Apr;7(4):509-30.

[6] Svidzinsky AA, Scully MO, Herschbach DR. Bohr's 1913 molecular model revisited. *Proc. Natl. Acad. Sci. U S A.* 2005 Aug 23;102(34):11985-8.

[7] Ben-Amotz D. Revisiting Bohr's semiclassical quantum theory. *J. Phys. Chem. B.* 2006 Oct 12;110(40):19861-6.

[8] Wiederkehr KH. Photoeffects, Einstein's light quanta and the history of their acceptance. *Sudhoffs Arch.* 2006;90(2):132-42.

[9] Graudenz K, Raulin C. From Einstein's Quantum Theory to novel laser therapy. The history of lasers in dermatology and aesthetic medicine] *Hautarzt.* 2003 Jul;54(7):575-82.

[10] Fischer-Hjalmars I. Molecular quantum mechanics in biology. *Q Rev. Biophys.* 1969 Jan;1(4):311-45.

[11] Burden SJ, McKay RD. Quantum mechanics of synapses. *Cell.* 1990 Oct 5;63(1):7-8.

[12] Conrad M. Quantum mechanics and cellular information processing: the self-assembly paradigm. *Biomed. Biochim. Acta.* 1990;49(8-9):743-55.

[13] Friesner RA, Beachy MD. Quantum mechanical calculations on biological systems. *Curr. Opin. Struct. Biol.* 1998 Apr;8(2):257-62.

[14] Gao J, Truhlar DG. Quantum mechanical methods for enzyme kinetics. *Annu. Rev. Phys. Chem.* 2002;53:467-505.

[15] Gorelsky SI, Lever ABP. Electronic structure and spectra of ruthenium diimine complexes by density functional Tteory and INDO/S. comparison of the two Methods. *J. Organomet. Chem.* 2001; 635: 187-196.

[16] Morokuma K. New challenges in quantum chemistry: Quests for accurate calculations for large molecular systems. *Philos. Transact. A Math. Phys. Eng. Sci.* 2002 Jun 15;360(1795):1149-64.

[17] Handoll N. Energy medicine: an osteopath's personal view. *J. Altern. Complement Med.* 2004 Feb;10(1):87-9.

[18] Hankey A. Are we close to a theory of energy medicine? *J. Altern. Complement Med.* 2004 Feb;10(1):83-6.

[19] Jain KK. Applied usages of Nanobiotechnology in Clinical Diagnostics. *Clin. Chem.* 2007 Sep 21;

[20] Hild WA, Breunig M, Goepferich A. Quantum dots - Nano-sized probes for the exploration of cellular and intracellular targeting. *Eur. J. Pharm. Biopharm.* 2007 Jun 15;

[21] Raymo FM, Yildiz I. Luminescent chemosensors based on semiconductor quantum dots. *Phys. Chem. Chem. Phys.* 2007 May 7;9(17):2036-43.

[22] Basabe-Desmonts L, Reinhoudt DN, Crego-Calama M. Design of fluorescent materials for chemical sensing. *Chem. Soc. Rev.* 2007 Jun;36(6):993-1017.

[23] Heuff RF, Swift JL, Cramb DT. Fluorescence correlation spectroscopy using quantum dots: advances, challenges and chances. *Phys. Chem. Chem. Phys.* 2007 Apr 28;9(16):1870-80.

[24] Pathak S, Davidson MC, Silva GA. Characterization of the functional binding properties of antibody conjugated quantum dots. *Nano Lett.* 2007 Jul;7(7):1839-45.

[25] Huo Q. A perspective on bioconjugated nanoparticles and quantum dots. *Colloids Surf. B Biointerfac*es. 2007 Sep 1;59(1):1-10.

[26] Pathak S, Davidson MC, Silva GA. Characterization of the functional binding properties of antibody conjugated quantum dots. *Nano Lett.* 2007 Jul;7(7):1839-45.

[27] Chan WC, Nie S. Quantum dot bioconjugates for ultrasensitive nonisotopic detection. *Science.* 1998 Sep 25;281(5385):2016-8.

[28] Goldman ER, Balighian ED, Mattoussi H, Kuno MK, Mauro JM, Tran PT, Anderson GP. Avidin: a natural bridge for quantum dot-antibody conjugates. *J. Am. Chem. Soc.* 2002 Jun 5;124(22):6378-82.

[29] Tokumasu F, Dvorak J. Development and applied usage of quantum dots for immunocytochemistry of human erythrocytes. *J. Microsc.* 2003 Sep;211(Pt 3):256-61.

[30] de Farias PM, Santos BS, de Menezes FD, de Carvalho Ferreira R, Barjas-Castro ML, Castro V, Lima PR, Fontes A, Cesar CL. Investigation of red blood cell antigens with highly fluorescent and stable semiconductor quantum dots. *J. Biomed. Opt.* 2005 Jul-Aug;10(4):44023.

[31] de Farias PM, Santos BS, de Menezes FD, Ferreira Rde C, Barjas-Castro Mde L, Castro V, Lima PR, Fontes A, Cesar CL. Core-shell CdS/Cd(OH)2 quantum dots: synthesis and bioconjugation to target red cells antigens. *J. Microsc.* 2005 Sep;219(Pt 3):103-8.

[32] Akhtar RS, Latham CB, Siniscalco D, Fuccio C, Roth KA. Immunohistochemical detection with quantum dots. *Methods Mol. Biol.* 2007;374:11-28.

[33] Ma Q, Wang X, Li Y, Su X, Jin Q. The use of CdTe quantum dot fluorescent microshperical objects in fluoro-immunoassays and a microfluidic chip system. *Luminescence.* 2007 Sep-Oct;22(5):438-45.

[34] Dwarakanatha S, Bruno JG, Athmaram TN, Bali G, Vattem D, Rao P. Antibody-quantum dot conjugates exhibit amplified antibacterial effect vs. unconjugated quantum dots. *Folia Microbiol.* (Praha). 2007;52(1):31-4.

[35] Shen J, Xu F, Jiang H, Wang Z, Tong J, Guo P, Ding S. Characterization and applied usage of quantum dot nanocrystal-monoclonal antibody conjugates for the determination of sulfamethazine in milk by fluoroimmunoassay. *Anal. Bioanal. Chem.* 2007 Sep 26;

[36] Wang HQ, Wang JH, Li YQ, Li XQ, Liu TC, Huang ZL, Zhao YD. Multi-color encoding of polystyrene microbeads with CdSe/ZnS quantum dots and its applied usage in immunoassay. *J. Colloid Interface Sci.* 2007 Sep 1;

[37] Liu G, Lin YY, Wang J, Wu H, Wai CM, Lin Y. Disposable electrochemical immunosensor diagnosis device based on nanoparticle probe and immunochromatographic strip. *Anal Chem.* 2007 Oct 15;79(20):7644-53.

[38] Tomlinson ID, Mason JN, Blakely RD, Rosenthal SJ. Peptide-conjugated quantum dots: imaging the angiotensin type 1 receptor in living cells. *Methods Mol. Biol.* 2005;303:51-60.

[39] Young SH, Rozengurt E. Qdot nanocrystal conjugates conjugated to bombesin or ANG II label the cognate G protein-coupled receptor in living cells. *Am. J. Physiol. Cell Physiol.* 2006 Mar;290(3):C728-32.

[40] Tsay JM, Trzoss M, Shi L, Kong X, Selke M, Jung ME, Weiss S. Singlet oxygen production by Peptide-coated quantum dot-photosensitizer conjugates. *J. Am. Chem. Soc.* 2007 May 30;129(21):6865-71.

[41] Medintz IL, Uyeda HT, Goldman ER, Mattoussi H. Quantum dot bioconjugates for imaging, labelling and sensing. *Nat. Mater.* 2005 Jun;4(6):435-46.

[42] Michalet X, Pinaud FF, Bentolila LA, Tsay JM, Doose S, Li JJ, Sundaresan G, Wu AM, Gambhir SS, Weiss S. Quantum dots for live cells, in vivo imaging, and diagnostics. *Science.* 2005 Jan 28;307(5709):538-44.

[43] Arya H, Kaul Z, Wadhwa R, Taira K, Hirano T, Kaul SC. Quantum dots in bio-imaging: Revolution by the small. *Biochem. Biophys. Res. Commun.* 2005 Apr 22;329(4):1173-7.

[44] Lidke DS, Arndt-Jovin DJ. Imaging takes a quantum leap. *Physiology* (Bethesda). 2004 Dec;19:322-5.

[45] Smith AM, Nie S. Semiconductor quantum dots as biological imaging agents. *Angew Chem. Int. Ed. Engl.* 2004 Aug 13;43(32):4129-31.

[46] Jiang W, Papa E, Fischer H, Mardyani S, Chan WC. Semiconductor quantum dots as contrast agents for whole animal imaging. *Potentials Biotechnol.* 2004 Dec;22(12):607-9.

[47] Jaiswal JK, Simon SM. Potentials and pitfalls of fluorescent quantum dots for biological imaging. *Potentials Cell Biol.* 2004 Sep;14(9):497-504.

[48] Kawasaki ES, Player A. Nanotechnology, nanomedicine, and the development of new, effective therapies for cancer. *Nanomedicine.* 2005 Jun;1(2):101-9.

[49] Joyner JC, Purcell WP. Quantum pharmacology and quantitative structure-activity relationships: a brief review. *Experientia Suppl.* 1976;23:13-23.

[50] Khromov BM. Use of the optical quantum generators (lasers) in neurosurgery (literature survey). *Vopr Neirokhir.* 1974 Nov-Dec;00(6):50-3

[51] Khromov BM. Use of the optic quantum generators (lasers) in therapeutic medicine. *Klin Med.* (Mosk). 1974 Nov;52(11):9-14.

[52] Khromov BM. Use of the optic quantum generators (lasers) in otorhinolaryngology. *Vestn. Otorinolaringol.* 1974 Jan-Feb;(1):94-7.

[53] Moskalik KG, Kozlov AP, Akimov AA. Use of optical quantum generators (lasers) in oncology. *Vopr Onkol.* 1972;18(8):97-105.

[54] Khromov BM, Frygin NV. The use of optic quantum generators (lasers) in dermatology (a review of foreign literature). *Vestn Dermatol. Venerol.* 1969 Apr;43(4):14-21.

[55] Bogdanovich UIa, Karimov MG. Use of optical quantum generators (lasers) to treat orthopedo-traumatologic patients. *Ortop. Travmatol .Protez.* 1984 Oct;(10):60-5.

[56] Karandashov VI, Petukhov EB. Quantum hemotherapy (history of the development of the method, its mechanism of action and prospects for its use in dentistry). *Stomatologiia* (Mosk). 1985 Jan-Feb;64(1):84-6.

[57] Moan J. Light therapy of newborns with hyperbilirubinemia. *Tidsskr. Nor. Laegeforen.* 1992 Apr 30;112(11):1459-63.

[58] Pratesi R, Agati G, Fusi F. Phototherapy for neonatal hyperbilirubinemia. *Photodermato*l. 1989 Dec;6(6):244-57.

[59] Wiwanitkit V. Quantum chemical analysis of alternative pathways for iron activation step for artemisinin, a new antimalarial medicine. *J. Infect.* 2006 Sep;53(3):148-51.

[60] Wiwanitkit V. Quantum chemical analysis of the deferiprone-iron binding reaction. *Int. J. Nanomedicine.* 2006;1(1):111-3.

[61] Wiwanitkit V. Analysis of binding energy activity of TIBO and HIV-RT based on simple consideration for conformational change. *Afr. J. Biotechnol.* 2007 Feb; 6(3): 188-189.

[62] Samia AC, Dayal S, Burda C. Quantum dot-based energy transfer: perspectives and potential for applied usages in photodynamic therapy. *Photochem. Photobiol.* 2006 May-Jun;82(3):617-25.

[63] Samia AC, Chen X, Burda C. Semiconductor quantum dots for photodynamic therapy. *J Am Chem Soc.* 2003 Dec 24;125(51):15736-7.

[64] Gorman A, Killoran J, O'Shea C, Kenna T, Gallagher WM, O'Shea DF. In vitro demonstration of the heavy-atom effect for photodynamic therapy. *J. Am. Chem. Soc.* 2004 Sep 1;126(34):10619-31.

[65] Schneider R, Tirand L, Frochot C, Vanderesse R, Thomas N, Gravier J, Guillemin F, Barberi-Heyob M. Recent improvements in the use of synthetic peptides for a selective photodynamic therapy. *Anticancer Agents Med. Chem.* 2006 Sep;6(5):469-88.

[66] Hsieh JM, Ho ML, Wu PW, Chou PT, Tsai TT, Chi Y. Iridium-complex modified CdSe/ZnS quantum dots; a conceptual design for bi-functionality toward imaging and photosensitization. *Chem. Commun* (Camb). 2006 Feb 14;(6):615-7.

[67] Upadhyay P. Amplified transdermal-immunization with diptheria-toxoid using local hyperthermia. *Vaccine.* 2006 Jul 7;24(27-28):5593-8.

[68] Ovcharova AP. The quantum gravitational therapy of myocarditis. *Lik Sprava.* 1999 Apr-May;(3):95-7.

[69] Kosov VA, Khuziev BG, Shchegol'kov AM, Nagatkin AN, Grubal'skaia IG. Spectral quantum therapy as a part of resort's rehabilitation program for patients with myocardial infarction. *Vopr. Kurortol. Fizioter. Lech. Fiz. Kult.* 2001 Jul-Aug;(4):9-13.

[70] Hossu M, Rupert R. Quantum events of biophoton emission associated with complementary and alternative medicine therapies: a descriptive pilot study. *J. Altern. Complement Med.* 2006 Mar;12(2):119-24.

Chapter 16

PICOMEDICINE: THE NEXT STEP

WHAT IS "PICOSCIENCE"?

As defined in the first chapter, nanolevel involves the range of size between 1 and 100 nm. However, in nature, there are many things that are below the 1 nm level. In the unit system, there are prefixes used to define objects below 1 nm. The generalized used prefixes are "pico" and "femto." "Pico" is the prefix describing the level of 10^{-3} Nm (10^{-9} m) or equal to 10^{-12} m. Many single molecules are in the picolevel. A good sample is a hydro (H_2O) molecule, which has a size equal to 10^{-1} nm or 10^{2} pm. In volume sense, many substances also exist in the picolevel, such as vitamins. Picoscience is a term used for describing the science focusing on objects that fall in the range between 1 and 100 pm. This science is believed to be the next step for nanoscience.

Similar to nanomaterial, picomaterial also possesses special properties that cannot be detected in general conditions. The quantum effect, surface effect, and interface effect can still be detected. In addition, visual comparison in judging the penetration via membrane layer can also be used. Since picomaterial is smaller than nanomaterial, the problem of permeability is very small.

REDOX IN MEDICINE

Oxidation and reduction are basic reactions in biological systems. The so-called term "oxidative stress" is commonly a focus in present medical research. Antioxidants have become a new magic tool for therapy of many complicated medical conditions. Indeed, an important change in redox reaction is the change of electron status between carrier and recipient. This phenomenon follows the general quantum medicine principle. (Please see the chapter on quantum medicine for more details.) The explanation of some biological redox reactions can be completely performed based on the quantum medicine reaction. To help the reader better understand the phenomenon, an example on the topic of reaction between nitric oxide and hemoglobin in the thalassemic patient (implication on hypertension) will be further shown.

INTRODUCTION

Nitric oxide is an important substance in humans. It appears to be involved in cellular defense against the iron-mediated reactive oxygen species generation probably mainly by inducing iron removal from cells [1]. In addition, NO may attenuate the effect of superoxide by mutual reaction, although the reaction product – peroxynitrite – is capable of producing highly reactive hydroxyl radicals [1]. NO is believed to reduce blood pressure by evoking vasodilation either directly by causing relaxation of vascular smooth muscle or indirectly by acting in the rostral brainstem to reduce central sympathetic outflow, which decreases the release of norepinephrine from sympathetic nerve terminals [2]. An increasingly large body of literature suggests that alterations in the NO system may play an important role in the development or maintenance of clinical hypertension [2].

Human endothelial nitric oxide synthase (eNOS) gene is responsible for production of NO and regulation of blood pressure [3]. A defect of the NO system is believed to be an important underlying condition for hypertension. In thalassemia, a common genetic disorder, hypertension can be seen. The most severe form, called syndrome of hypertension, convulsion, and cerebral hemorrhage (HCCH) can lead to fatality [4]. The main cause of this syndrome is not well-documented. Disturbance of the NO mechanism is believed to be an important cause.

Recently, it was proposed that hemolysis in thalassemic patients could bring disturbance in NO metabolism and lead to hypertension. Reddy et al found that an association between alpha-thalassemia and hypertension might involve trapping of the vasodilator NO by hemoglobin (Hb) [5]. Here, the author analyzes the reaction between NO and oxyhemoglobin based on the basic principle of quantum chemical analysis, and implies a correlation to hypertension.

MATERIALS AND METHODS

Pathways for reaction between oxyhemoglobin and nitric oxide is hereby described. Concerning the reaction between oxyhemoglobin (Hb-O2) and nitric oxide (NO), the reaction will bring methemoglobin (MetHb) and oxygen (O2) as results. This pathway is irreversible. The quantum chemical analysis for total reaction was performed according to the classical bonding theory [6]. The required energy for complex formation in each pathway was calculated.

RESULTS

The details and the required energy for the reaction are presented in Table 1. The required energy for reaction is – 1 eV.

Table 1. The details and the required energy in each step of reaction

Items	Details
Bond breaking*	1 Hb-O2
Bond forming**	1 Hb-NO
Obtainting in energy	1 eV
Giving out energy	2 eV
Required energy***	- 1 eV

*bond breaking obtainting in energy
**bond forming giving out energy
***required energy = obtainting in energy - giving out energy

DISCUSSION

Although NO is considered a free radical, it is relatively stable. It reacts predominantly with molecules that have orbitals with unpaired electrons, which are typically other free radicals or transition metals like hemoglobin. This reaction is believed to be a major underlying germesis for hypertension in thalassemia, which has the high rate of hemolysis and hemoglobin turnover [7].

The bonding change in this reaction is according to the redox reaction principle. Here, the author calculated the bonding energy of the reaction and found that the reaction is a type of "energy producing reaction". This finding implies many in clinical medicine. It means that the reaction can occur spontaneously without an external source of energy. This finding implies that the reaction between Hb-O2 and NO can occur in vivo with the supplement of Hb from hemolytic episode in thalassemia. Hence, if the hemolysis increases, the reaction will increase and result in the reduction of NO. Indeed, intravascular hemolysis releases hemoglobin from the red blood cell into the plasma (cell-free plasma hemoglobin), which is then able to scavenge endothelium-derived NO 600-fold faster than erythrocytic hemoglobin, thereby disrupting NO homeostasis [8]. The degree of hemolysis is also reported to relate to decrease of NO level [8].

In addition, the nature of an energy-producing reaction implies the supplying of energy to the nearby external environment. Endothelial cells can be impacted by the derived energy. This result is concordant with the recent knowledge that endothelial dysfunction, vasomotor tone and organ perfusion disturbance might be impaired by the increased reactivity of cell-free plasma Hb with NO during intravascular hemolysis [9]. Indeed, the pathology of endothelium in thalassemia is warranted [10]. The endothelium defect might also bring a defect in production of NO as a consequence, and this can also lead to hypertension. Conclusively, the author proposes that the nature of an energy-producing reaction in a reaction between Hb-O2 and NO is a main underlying cause for the hypertension in thalassemic patients.

There are some reports on the redox in picomedicine. The details are presented in Table 2.

Table 2. Some reports on the redox in picomedicine

Authors	Details
Maret [11]	Maret studied how cellular zinc and redox states converge in the metallothionein/thionein pair [11]. Maret proposed that selenium expressed its antioxidant effects through redox catalysis in zinc metabolism [11].
Calcaterra et al. [12]	The contribution made by tyrosine 308 to the stability of pea ferredoxin-NADP+ reductase was investigated using site-directed mutagenesis [12]. Calcaterra et al. indicated that the existence of the electron-rich aromatic side chain adjacent to the isoalloxazine ring was essential for maximum stabilization of the FNR holoenzyme, resulting in a flavin conformation which optimized electron flow between the prosthetic group and its redox partners [12].
Katou et al. [13]	Katou et al. performed this work to clarify the role of the disulfide bond in amyloid fibril formation. They characterized the conformations of the oxidized and reduced forms of beta2-m in the acid-denatured state at pH 2.5, as well as the native state at pH 6.5, by heteronuclear NMR. [(1)H]-(15)N NOE at the regions between the two cysteine residues (Cys25-Cys80) revealed a marked difference in the pico- and nanosecond time scale dynamics between that the acid-denatured oxidized and reduced states, with the former showing reduced mobility [13].
Vakrat-Haglili et al. [14]	Vakrat-Haglili et al. revealed that Pd-bacteriopheophorbide, a useful reagent for vascular targeted PDT of solid tumors, which had recently entered into phase II clinical experiments under the code name WST09, forms appreciable summations of hydroxyl radicals, superoxide radicals, and probably hydrogen peroxide in aqueous medium but not in organic solvents where singlet oxygen almost exclusively forms [14].
Berneis [15]	Berneis reported that catalysis and inhibition of the oxidative degradation of deoxyribonucleic acid by catalase revealed maximum catalytic effect in pico-molar concentrations [15].
Horne et al. [16]	Horne et al. detected the structure of the reduced form of the DsbA oxidoreductase from *Vibrio cholerae* [16]. In this work, the reduced structure revealed a high level of similarity to the crystal structure of the oxidized form and is typical of this class of enzyme consist ofing a thioredoxin domain with an inserted alpha-helical domain [16]. Horne et al. also mentioned that proteolytic and thermal stability measurements revealed that the reduced form of DsbA was considerably more stable than the oxidized form [16].

PICOQUANTITY MEASUREMENT IN MEDICINE

Similar to nanoquantity measurement in medicine, picoquantity measurement in medicine faces similar problems. Determination for very low quantity has to be done with a high analytical sensitivity method. The presently used technique in medical laboratory practice is electrochemiluminescence [16 – 20] (please see the chapter on nanoquantity measurement in medicine for more details). The main pico-quantity substances that are determined in laboratory medicine are vitamins and hormones. Some interesting reports on picoquantity measurement in medicine are presented in Table 3.

Table 3. Reports on pico-quantity measurement in medicine

Authors	Details
Yan et al. [21]	In this work, a protocol for quantitative detection of p53 antibodies in human serum using immunomagnetic electrochemiluminescence was developed [21]. The assay uses only 50 microl of sample per test and requires a 30-min incubation period in addition to a 50 s acquisition time [21]. This assay has many usefulness over the commonly used ELISA method in terms of sensitivity, linear range, and assay time [21]. Yan et al. indicated that the sensitivity of this assay was 10 pg of p53 antibodies per ml of reference serum (normal human serum) [21].
Guglielmo-Viret and Thullier [22]	In this work, an electrochemiluminescence assay for the detection of the B chain of ricin (RCA-B) in a 96-well plate format was developed in parallel with a colorimetric ELISA utilizing the same pair of antibodies [22]. According to this work, the electrochemiluminescence assay was also three times faster, and retained the robustness and convenience of the 96-well plate format [22].
So et al. [23]	In this work, simultaneous optical and voltammetric detection of bioactivated genotoxicity is reported for the first time employing ultrathin films of DNA, model metabolic enzymes, and electrochemiluminescence generating metallopolymer [Ru(bpy)(2)PVP(10)](2+) on pyrolytic graphite (PG) electrodes [23].
Poli et al. [24]	A new competitive electrochemiluminescence-based immunoassay for the type-2 brevetoxins in oyster extracts was developed [24]. Poli et al. believed this assay could be useful as a regulatory screening tool and could support pharmacokinetic studies in animals and clinical evaluation of neurotoxic shellfish poisoning victims [24]. Poli et al. said that low limit of quantitation (50 pg/mL) could be detected [24].
Rivera et al. [25]	In this work, sensitive and specific electrochemiluminescence assays were used to detect *Clostridium botulinum* neurotoxins serotypes A, B, E, and F in undiluted human serum, undiluted human urine, assay buffer, and selected food matrices (whole milk, apple juice, ground beef, pastry, and raw eggs) [25]. Rivera et al. said that the assay might have potential use in clinical settings, research studies, and screening of food products for botulinum toxins [25].
Golla and Seethala [26]	Golla and Seethala reported a robust, extremely sensitive electrochemiluminescence insulin assay using the Origen technology platform [26]. According to this work, the dynamic range of detection for insulin is 5 pg to 5 ng in the electrochemiluminescence assays [26]. Golla and Seethala suggested that the ECL insulin assay was an extremely sensitive, robust, nonradioactive homogeneous assay and could be used completely to determine the insulin levels in rodent serum samples [26].
Sennikov et al. [27]	Sennikov et al. described the quantification of interleukin (IL)-2, IL-4, IL-10 and interferon-gamma in human serum and peripheral blood mononuclear cell-conditioned medium by electrochemiluminescence method [27]. Sennikov et al. demonstrated that this approach allowed one to detect cytokine concentration from 1 pg/ml [27].
Arai et al. [28]	In this work, an electrochemiluminescence flow-through cell with a carbon fiber electrode for electrochemiluminescence detection was developed, and a flow injection analysis system consist ofing the cell was established. N-(aminobutyl)-N-ethylisoluminol, an electrochemiluminescence reagent, was detected by this system [28].

PICOFLUIDICS

Picofluidics is the newest generation of fluidics technology. Starting from microfluidics, in which a small device with micrometer size controls the flow of fluid (liquid and gas), nanofluidics was developed. Nanofluidics is a similar device that involves nanometer-size compositions (such as nanotubes), which is commonly used as chip technology for point-of-care testing. Jumping into the third generation of fluidics, picofluidics has been in existence for only a few years. PicoLiter (USA) is an example of a company that performed research and development in the area of picofluidics.

In addition to picofluidics, the picobiosensor is another new hope. Also, there are many examples of current research to develop a pico-level biosensor. The platform under development, namely "Pico Titer sequencing plate" by 454 Life Science (USA), is a good example of research and development on a pico-level biosensor.

Table 4. Some reports on picofluidics

Authors	Details
Beer et al. [29]	In this work, on-chip, real-time, single-copy polymerase chain reaction (PCR) in picoliter droplets was reported [29]. With this system, a 10-pL droplet, encapsulating less than one copy of viral genomic DNA through Poisson statistics, revealed real-time PCR amplification curves with a cycle threshold of approximately 18, 20 cycles earlier than commercial tools [29].
Srisa-Art et al. [30]	High-throughput DNA droplet assays using picoliter reactor volumes was reported [30]. Srisa-Art et al. described the integration of confocal fluorescence spectroscopy as a high-efficiency detection method for droplet-based microfluidics [30]. In this work, issues such as surface contamination, rapid mixing, and rapid detection, too as low detection limits was addressed with the approach described when compared to conventional laminar flow-based fluidics [30].
Lee et al. [31]	In this work, Lee et al. presented a new way to transport and handle picoliter volumes of analytes in a microfluidic context through electrically monitored electroporation of 10-25 microm vesicles [31]. This work demonstrated the probability of using giant, robust vesicles with microfluidic electroporation technology to manipulate picoliter volumes on-chip [31].
Irimia et al. [32]	Irimia et al. proposed a microfabricated device for the controlled handling and mixing of picoliter cell suspension and lysis solution volumes [32]. In this work, cells and fluids were independently isolated in two microchambers of 25-pL volumes using the geometry of the microchannels and the coordinated action of four on-chip thermopneumatic actuators [32].
Marcus et al. [33]	Marcus et al. developed a fluidic chip to perform 72 parallel 450-pL RT-PCRs [33]. They reported that the device and method presented might enable highly parallel single cell gene expression analysis [33].

PICODIAGNOSIS AND PICODEVICE

Similar to nanodiagnosis, picodiagnosis can be performed. Based on the property of very small objects, applied usage of pico-sized materials can be used for picodiagnosis. At present, picoarray isavailable. In 2004, a picoliter chamber array for cell-free protein synthesis was reported by Kinpara et al. [33]. In this study, Kinpara et al. attempted to make a highly-integrated protein chip from a DNA library using in vitro protein synthesis on a microchamber array fabricated by using PDMS (polydimethyl siloxane), a hydrophobic surface, and glass, a hydrophilic bottom substrate [34]. Kinpara et al. reported that these structural properties prevented cross-contamination among the chambers and the minimum volume capacity of the smallest chamber was about 1 pL [34]. In 2004, Zhou et al. described a PicoArray, which involves a massively parallel array of four different kinds of processors, a method for the simultaneous synthesis and purification of oligonucleotides that are designed for multiplex gene synthesis [35]. Zhou et al. illustrated the potential of fluidics-based ultra-fast oligonucleotide parallel synthesis as an enabling tool for novel synthetic biology applied usages, such as the construction of genome-scale molecular clones and cell-free large-scale protein expression [35].

As with nanotechnology, picotechnology enables a new revolution. At present, due to the advent in picoelectronics, there are some available picodevices. These picodevices can be useful for picodiagnosis in the future. It is belived that "quantum machinery" will soon be able to work with subatomic particles and quantum field events. More details on quantum medicine for diagnosis can be seen in another chapter in this book.

Table 5. Some reports on picodevices

Authors	Details
Risveden et al. [36]	In this work, a new type of bioelectronic region ion-sensitive field effect transistor nanosensor was constructed [36]. For glucose measurements, negatively charged gluconate ions were gathered between the sensing electrodes and the signal present response was determined using a low-noise pico ammeter [36]. The sensitivity was 28 pA/mM and 830 pA/mM, respectively [36]. Risveden et al. reported that the volumes expressed in this way could be in low or sub femtoliter range [36].
Simson et al. [37]	In-depth analysis and experimental verification on micropipet-based pico force transducer was reported [37].
Votaw and Shulman [38]	Votaw and Shulman assessd the Pico-Count flow-through radioactivity detector, designed for use in positon emission tomography (PET) studies of cerebral blood flow [38]. Votaw and Shulman concluded that the Pico-Count was an excellent detector for continuously monitoring positron radioactivity in blood [38]. Votaw and Shulman also summarized that dispersion correction for the detection apparatus might not be needed depending on the applied usage [38].

PICO IMAGING

Pico imaging is available at present. Quantum dot imaging can be classified as an applied usage for pico imaging. However, the best commonly used pico imaging is the PET scan. A positon is a particle that can be easily described as a positive charge electron. It is an antiparticle or antimatter counterpart of the electron. There are many reports on pico imaging and PET scan as presented in Table 6.

Table 6. Some reports on pico imaging and PET

Authors	Details
Matsumoto et al. [39].	Matsumoto et al. examined the characteristic of three different types of continuous blood sampling systems: a plastic scintillator-based beta detector [conventional beta detector (BETA)], a bismuth germanate (BGO)-based coincidence gamma detector [Pico-count flow-through detector (COINC)] and a Phoswich detector (PD) composed by a combination of plastic scintillator and BGO scintillator [39]. Matsumoto et al. concluded that although the characteristic of all three sensors meets clinical requirements, PD had the highest physical characteristic [39].
Seemann et al. [40]	This prospective study included patients with extensive nonresectable metastases of gastrointestinal carcinoid tumors which were consecutively examined from the base of the skull to the proximal thigh using a state-of-the-art PET/CT scanner and a 1.5 Tesla whole-body MRI scanner [40]. PET was performed with a carbohydrated F-18-labeled somatostatin-receptor ligand ([superset18 F]FP-Gluc-TOCA) using a Pico-3D PET scanner [40].
Seemann [41]	Prospective comparison of 18F-TOCA PET, triple-phase CT, and PET/CT for detection of metastases from gastrointestinal neuroendocrine tumors was performed [41]. In this work, PET was performed with a carbohydrated F-18-labeled somatostatin receptor ligand (Gluc-Lys([(18)F]FP-TOCA) from the base of the skull to the proximal thigh using a Pico-3D PET scanner [41].
Martinez et al. [42]	Martinez et al. assessd the characteristic of the BS16 PET scanner before and after the Pico-3D electronics upgrade [42]. Martinez et al. found that the characteristic of the PET scanners in the BS16 tomographs was improved by the optimization of the signal processing and the narrower energy and timing coincidence windows led to a considerable increase of signal- to-noise ratio [42].

Another well-known method for pico imaging is the atomic force microscope [43 – 45]. The sensitivity of the AFM cantilever, to forces in the pico Newton range, has been exploited to measure breakaway forces between biomolecules and to measure folding-unfolding forces within single proteins [43]. Hembacher et al. said that a low-temperature atomic force microscope with pico-Newton force sensitivity could help reveal the hidden surface atom [44]. More details on the atomic force microscope can be seen in another chapter in this book.

PICOTHERAPY

Picotherapy is a new concept that has not been realized at present. Picotherapy is believed to be a highly effective tool for treatment of many untreatable diseases. However, at present, nanotherapy is only in experiments; therefore, there is a very long road ahead for picotherapy. The applied usage of human genetic engineering or prenatal gene therapy might be the first to be completed. At present, genome Sequencer 20 System, a picotechnology-based genome sequencing system, has been completely developed, and this might be the first step for further gene therapy based on picobiotechnology.

FEMTOMEDICINE: CAN WE REACH ?

It is accepted that picotechnology is the new next step for nanotechnology. However, the process is continuously evolving. The next step to femtomedicine is a distant goal. The main question is we can reach femtotechnology. Femtotechnology involves objects which fall in the range between 1 and 100 pm, which are extremely small. Femtotechnology can involve the manipulation of excited energy states within atomic nuclei. This question remains a challenging one in science. Indeed, at present, scientists can determine the femto volume quantity. For example, in 2005, He et al. described a method that combines optical trapping and fluidic-based droplet generation, for selectively and controllably encapsulating a single target cell or subcellular structure, such as a mitochondrion, into a picoliter- or femtoliter-volume aqueous droplet that is surrounded by an immiscible phase [46]. There are also some reports on the femtoscale in medicine. The femtosecond laser (Femto-plak) is the best example. In ophthalmology, femtosecond laser is presently used for corneal endothelial decompensation [47]. Another example is the ability to detect the femto volume of a biological analysis in laboratory medicine by electrochemiluminesence method [48 - 49]. In 2006, Scherer et al. demonstrated pico-liter sample delivery and femto-liter analytical volumes [50]. This might be the first phase of the development of femtofludics. In addition, there was a recent meeting in 2007, Femto Forum [51] to explore the guidelines and standards for femtotechnology. Femto Forum aims to bring femtocell technology worldwide. Although this is for engineering, it can be a first step for bio-applied usage. It can be said that femtomedicine can be the next step to be reached.

REFERENCES

[1] Mladenka P, Simunek T, Hubl M, Hrdina R. The role of reactive oxygen and nitrogen species in cellular iron metabolism. *Free Radic. Res*. 2006;40:263-72.

[2] Augustyniak RA, Thomas GD, Victor RG, Zhang W. Nitric oxide pathway as new medicine targets for refractory hypertension. *Curr. Pharm. Des*. 2005;11:3307-15.

[3] Wang X, Cade R, Sun Z. Expression of human eNOS in cardiac and endothelial cells. *Methods Mol. Med.* 2005;112:91-107.

[4] Wiwanitkit V. Brain pathology in a syndrome of hypertension, convulsion, and cerebral haemorrhage in thalassaemic patients after multiple blood-transfusions: a summary in reported Thai autopsy cases. *J. Hypertens.* 2006;24:601.

[5] Reddy PL, Bowie LJ, Callistein S. Binding of nitric oxide to thiols and hemes in hemoglobin H: implications for alpha-thalassemia and hypertension. *Clin. Chem.* 1997;43(8 Pt 1):1442-7

[6] Oldfield E. Quantum chemical studies of protein structure. *Philos. Trans. R. Soc. Lond B Biol. Sci.* 2005;360:1347-61.

[7] Schrier SL. Pathophysiology of thalassemia. *Curr. Opin. Hematol.* 2002;9:123-6.

[8] Gladwin MT, Crawford JH, Patel RP. The biochemistry of nitric oxide, nitrite, and hemoglobin: role in blood flow regulation. *Free Radic. Biol. Med.* 2004;36:707-17.

[9] Minneci PC, Deans KJ, Zhi H, Yuen PS, Star RA, Banks SM, Schechter AN, Natanson C, Gladwin MT, Solomon SB. Hemolysis-associated endothelial dysfunction mediated by accelerated NO inactivation by decompartmentalized oxyhemoglobin. *J. Clin. Invest.* 2005;115:3409-17.

[10] Butthep P, Nuchprayoon I, Futrakul N. Endothelial injury and altered hemodynamics in thalassemia. *Clin. Hemorheol. Microcirc.* 2004;31:287-93.

[11] Maret W. Cellular zinc and redox states converge in the metallothionein/thionein pair. *J. Nutr.* 2003 May;133(5 Suppl 1):1460S-2S.

[12] Calcaterra NB, Pico GA, Orellano EG, Ottado J, Carrillo N, Ceccarelli EA. Contribution of the FAD binding site residue tyrosine 308 to the stability of pea ferredoxin-NADP+ oxidoreductase. *Biochemistry.* 1995 Oct 3;34(39):12842-8.

[13] Katou H, Kanno T, Hoshino M, Hagihara Y, Tanaka H, Kawai T, Hasegawa K, Naiki H, Goto Y. The role of disulfide bond in the amyloidogenic state of beta(2)-microglobulin studied by heteronuclear NMR. *Protein Sci.* 2002 Sep;11(9):2218-29.

[14] Vakrat-Haglili Y, Weiner L, Brumfeld V, Brandis A, Salomon Y, McLlroy B, Wilson BC, Pawlak A, Rozanowska M, Sarna T, Scherz A. The microenvironment effect on the generation of reactive oxygen species by Pd-bacteriopheophorbide. *J. Am. Chem. Soc.* 2005 May 4;127(17):6487-97.

[15] Berneis K. Catalysis and inhibition of the oxidative degradation of deoxyribonucleic acid by catalase showing maximum catalytic effect in pico-molar concentrations. *Experientia.* 1967 Feb 15;23(2):91-2.

[16] Electrochemical detection in bioanalysis. *J. Pharm. Biomed. Anal.* 1996;14:871-80.

[17] Richter MM. Electrochemiluminescence (ECL). Chem Rev. 2004;104:3003-36.

[18] Bruce D, McCall J, Richter MM. Effects of electron withdrawing and donating groups on the efficiency of tris(2,2'-bipyridyl)ruthenium(II)/tri-n-propylamine electrochemiluminescence. *Analyst.* 2002;127:125-8.

[19] Amatore C, Pebay C, Servant L, Sojic N, Szunerits S, Thouin L. Mapping electrochemiluminescence as generated at double-band microelectrodes by confocal microscopy under steady state. *Chemphyschem.* 2006;7:1322-7.

[20] Calcaterra NB, Pico GA, Orellano EG, Ottado J, Carrillo N, Ceccarelli EA. Contribution of the FAD binding site residue tyrosine 308 to the stability of pea ferredoxin-NADP+ oxidoreductase. *Biochemistry.* 1995 Oct 3;34(39):12842-8.

[21] Yan G, Xing D, Tan S, Chen Q. Rapid and sensitive immunomagnetic-electrochemiluminescent detection of p53 antibodies in human serum. *J. Immunol. Methods.* 2004 May;288(1-2):47-54.

[22] Guglielmo-Viret V, Thullier P. Comparison of an electrochemiluminescence assay in plate format over a colorimetric ELISA, for the detection of ricin B chain (RCA-B). *J Immunol Methods.* 2007 Aug 29;

[23] So M, Hvastkovs EG, Schenkman JB, Rusling JF. Electrochemiluminescent /voltammetric toxicity screening sensor using enzyme-generated DNA damage. *Biosens. Bioelectron.* 2007 Nov 30;23(4):492-8.

[24] Poli MA, Rivera VR, Neal DD, Baden DG, Messer SA, Plakas SM, Dickey RW, Said KE, Flewelling L, Green D, White J. An electrochemiluminescence-based competitive displacement immunoassay for the type-2 brevetoxins in oyster extracts. *J. AOAC Int.* 2007 Jan-Feb;90(1):173-8.

[25] Rivera VR, Gamez FJ, Keener WK, White JA, Poli MA. Rapid detection of Clostridium botulinum toxins A, B, E, and F in clinical samples, selected food matrices, and buffer using paramagnetic bead-based electrochemiluminescence detection. *Anal. Biochem.* 2006 Jun 15;353(2):248-56.

[26] Golla R, Seethala R. A sensitive, robust high-throughput electrochemiluminescence assay for rat insulin. *J. Biomol .Screen.* 2004 Feb;9(1):62-70.

[27] Sennikov SV, Krysov SV, Injelevskaya TV, Silkov AN, Grishina LV, Kozlov VA. Quantitative analysis of human immunoregulatory cytokines by electrochemiluminescence method. *J. Immunol. Methods.* 2003 Apr 1;275(1-2):81-8.

[28] Arai K, Takahashi K, Kusu F. An electrochemiluminescence flow-through cell and its applied usages to sensitive immunoassay using N-(aminobutyl)-N-ethylisoluminol. *Anal. Chem.* 1999 Jun 1;71(11):2237-40.

[29] Beer NR, Hindson BJ, Wheeler EK, Hall SB, Rose KA, Kennedy IM, Colston BW. On-Chip, Real-Time, Single-Copy Polymerase Chain Reaction in Picoliter Droplets. *Anal. Chem.* 2007 Oct 11;

[30] Srisa-Art M, deMello AJ, Edel JB. High-throughput DNA droplet assays using picoliter reactor volumes. *Anal. Chem.* 2007 Sep 1;79(17):6682-9.

[31] Lee ES, Robinson D, Rognlien JL, Harnett CK, Simmons BA, Bowe Ellis CR, Davalos RV. Microfluidic electroporation of robust 10-microm vesicles for manipulation of picoliter volumes. *Bioelectrochemistry.* 2006 Sep;69(1):117-25.

[32] Irimia D, Tompkins RG, Toner M. Single-cell chemical lysis in picoliter-scale closed volumes using a microfabricated device. *Anal Chem.* 2004 Oct 15;76(20):6137-43.

[33] Marcus JS, Anderson WF, Quake SR. Parallel picoliter rt-PCR assays using microfluidics. *Anal. Chem.* 2006 Feb 1;78(3):956-8.

[34] Kinpara T, Mizuno R, Murakami Y, Kobayashi M, Yamaura S, Hasan Q, Morita Y, Nakano H, Yamane T, Tamiya E. A picoliter chamber array for cell-free protein synthesis. *J. Biochem.* (Tokyo). 2004 Aug;136(2):149-54.

[35] Zhou X, Cai S, Hong A, You Q, Yu P, Sheng N, Srivannavit O, Muranjan S, Rouillard JM, Xia Y, Zhang X, Xiang Q, Ganesh R, Zhu Q, Matejko A, Gulari E, Gao X. Microfluidic PicoArray synthesis of oligodeoxynucleotides and simultaneous assembling of multiple DNA sequences. *Nucleic. Acids. Res.* 2004 Oct 11;32(18):5409-17.

[36] Risveden K, Ponten JF, Calander N, Willander M, Danielsson B. The region ion sensitive field effect transistor, a new bioelectronic nanosensor. *Biosens Bioelectron.* 2007 Jun 15;22(12):3105-12.

[37] Simson DA, Ziemann F, Strigl M, Merkel R. Micropipet-based pico force transducer: in depth analysis and experimental verification. *Biophys. J.* 1998 Apr;74(4):2080-8.

[38] Simson DA, Ziemann F, Strigl M, Merkel R. Micropipet-based pico force transducer: in depth analysis and experimental verification. *Biophys. J.* 1998 Apr;74(4):2080-8.

[39] Matsumoto K, Shinoda M, Yamamoto S, Tarutani K, Minato K, Sakamoto S, Senda M. Characteristic evaluation of continuous blood sampling system for PET study: comparison of three detector-systems. *Kaku Igaku.* 2002 Nov;39(4):527-34.

[40] Seemann MD, Meisetschlaeger G, Gaa J, Rummeny EJ. Assessment of the extent of metastases of gastrointestinal carcinoid tumors using whole-body PET, CT, MRI, PET/CT and PET/MRI. *Eur. J. Med. Res.* 2006 Feb 21;11(2):58-65.

[41] Seemann MD. Detection of metastases from gastrointestinal neuroendocrine tumors: prospective comparison of 18F-TOCA PET, triple-phase CT, and PET/CT. *Technol. Cancer Res. Treat.* 2007 Jun;6(3):213-20.

[42] Martinez MJ, Bercier Y, Schwaiger M, Ziegler SI. PET/CT Biograph Sensation 16. Characteristic improvement using faster electronics. *Nuklearmedizin.* 2006;45(3):126-33.

[43] Allison DP, Hinterdorfer P, Han W. Biomolecular force measurements and the atomic force microscope. *Curr. Opin. Biotechnol.* 2002 Feb;13(1):47-51.

[44] Hembacher S, Giessibl FJ, Mannhart J, Quate CF. Revealing the hidden atom in graphite by low-temperature atomic force microscopy. Proc Natl Acad Sci U S A. 2003 Oct 28;100(22):12539-42.

[45] Gadegaard N. Atomic force microscopy in biology: technology and ways. *Biotech Histochem.* 2006 Mar-Jun;81(2-3):87-97.

[46] He M, Edgar JS, Jeffries GD, Lorenz RM, Shelby JP, Chiu DT. Selective encapsulation of single cells and subcellular organelles into picoliter- and femtoliter-volume droplets. *Anal. Chem.* 2005 Mar 15;77(6):1539-44.

[47] Seitz B, Langenbucher A, Hofmann-Rummelt C, Schlotzer-Schrehardt U, Naumann GO. Nonmechanical posterior lamellar keratoplasty using the femtosecond laser (femto-plak) for corneal endothelial decompensation. *Am. J. Ophthalmol.* 2003 Oct;136(4):769-72.

[48] Ratkay-Traub I, Ferincz IE, Juhasz T, Kurtz RM, Krueger RR. First clinical results with the femtosecond neodynium-glass laser in refractive surgery. *J. Refract. Surg.* 2003 Mar-Apr;19(2):94-103.

[49] Roda A, Manetta AC, Portanti O, Mirasoli M, Guardigli M, Pasini P, Lelli R. A rapid and sensitive 384-well microtitre format chemiluminescent enzyme immunoassay for 19-nortestosterone. *Luminescence.* 2003 Mar-Apr;18(2):72-8.

[50] Scherer A, Kartalov E, Anderson F. Integrated opto-fluidics for biochemical applied usages. In: Microfluidics and Nanotechnology: The Atom 's the Limit. Presented in The 40th Western Regional Meeting (January 22-25, 2006).

[51] The Femto Forum. Available online at http://www.femtoforum.org.

INDEX

A

B

C

D

E

H

I

J

K

L

N

O

P

Q

R

T

U

V

W

X

Y

Z